Stochastic Theory and
Cascade Processes

series

MODERN ANALYTIC AND COMPUTATIONAL METHODS IN SCIENCE AND MATHEMATICS
A Group of Monographs and Advanced Textbooks

Editor: RICHARD BELLMAN, University of Southern California

MODERN ANALYTIC AND COMPUTATIONAL METHODS IN SCIENCE AND MATHEMATICS

MÉTHODES MODERNES D'ANALYSE ET DE COMPUTATION EN SCIENCE ET MATHÉMATIQUE

NEUE ANALYTISCHE UND NUMERISCHE METHODEN IN DER WISSENSCHAFT UND DER MATHEMATIK

НОВЫЕ АНАЛИТИЧЕСКИЕ И ВЫЧИСЛИТЕЛЬНЫЕ МЕТОЛЫ В НАУКЕ И МАТЕМАТИКЕ

Editor: RICHARD BELLMAN
University of Southern California

Stochastic Theory and Cascade Processes

by

S. KIDAMBI SRINIVASAN

Department of Mathematics
Indian Institute of Technology, Madras

AMERICAN ELSEVIER PUBLISHING COMPANY, INC.
NEW YORK 1969

American Elsevier Publishing Company, Inc.
52 Vanderbilt Avenue
New York, N.Y. 10017

Elsevier Publishing Company
Barking, Essex, England

Elsevier Publishing Company
335 Jan Van Galenstraat, P.O. Box 211
Amsterdam, The Netherlands

Standard Book Number 444–00051–8

Library of Congress Card Number 69–13068

Printed in The United States of America

Dedicated to

MY MOTHER

and

THE MEMORY OF MY FATHER

PREFACE

About thirty years ago Bhabha and Heitler, in their study of the cascade phenomena in cosmic ray showers, showed how stochastic methods can be applied to branching phenomena. In the same year, Yvon in an attempt to study the statistical properties of an assembly of molecules introduced certain correlation functions which have dominated the thoughts and attempts of statistical physicists. These two events mark the birth of a new branch of stochastic theory now well known as point processes. Though the precise formulation of the theory of such processes in rigorous mathematical language is fairly difficult, its phenomenological interpretation with specific reference to the underlying phenomena has had its record of success in the past two decades. Many techniques have been developed particularly by Bogoliubov, Uhlenbeck, Bellman and Harris, Janossy, Bhabha, Ramakrishnan, and Kendall and these have proved to be valuable in the study of physical phenomena ranging from classical kinetic theory to control theory and nervous systems. Although a number of monographs on various aspects of stochastic processes have been published, point processes have not been dealt with in their proper context in any of the publications with the exception of those by A. T. Bharucha-Reid and T. E. Harris*. The object of the present work is to present an account of the methods of stochastic point processes that are being widely used in branching and other types of phenomena. Since the processes themselves have been largely responsible for the development of the theory, a monograph which will give an adequate account of both the theory and the processes is expected to be a useful supplement to the monographs of Harris and Bharucha-Reid.

Chapter 1, which is introductory in nature, contains an account of the common features of stochastic processes with special reference to multiplicative processes. The general theory of point processes is discussed in Chapters 2 and 3. We then present some results relating to the extensions of the product density technique and multiple point processes. Chapters 5 and 6 contain an account of electromagnetic cascades. In these two Chapters we have paid considerable attention to the analytic methods of solution and their computational aspects leading to numerical results of the quantities that can be experimentally determined. Chapter 7 deals with the development

* The survey article of A. Ramakrishnan in the Encyclopedia of Physics (Volume III) deals with some aspects of point processes.

of extensive air showers while Chapter 8 describes the polarization cascades involving product density matrices. In the final chapter we have included a brief account of the stochastic theory of population growth from the viewpoint of point processes.

In the choice of the material for the monograph, I have been mainly motivated by those physical processes that have had an impact on the development of stochastic theory in general. I have deliberately avoided any discussion on neutron branching processes since they have been dealt with in detail by Bellman, Kalaba, and others in some of the earlier monographs in the series. I have also not discussed the utility of Monte Carlo methods in the cascade phenomena the success of which has been amply demonstrated in the case of electromagnetic cascades by Messell and his collaborators.

The reader is assumed to have a knowledge at the level of Feller's *Probability Theory*. Throughout the monograph, the viewpoint is that of an applied mathematician and I have not hesitated to resort to phenomenological methods and approximate solutions in preference to rigorous treatment and analytical method wherever it is advantageous to do so. Since the methods and tools are the common stock of an applied mathematician, it is hoped that this monograph will be useful for graduate students who wish to specialize in stochastic processes and statistical physics.

I am deeply indebted to my former teacher, Professor Alladi Ramakrishnan who introduced me to this subject and to Professor Richard Bellman who encouraged me to write this monograph.

Thanks are due to Professor R. Vasudevan and N. V. Koteswara Rao who have read through the entire manuscript and suggested various improvements in the text as well as the mode of presentation.

I appreciate the excellent manner in which R. Subramanian and S. Kumaraswamy have helped me in the preparation of the manuscript. I am also thankful to S. M. Rajasekharan for typing the manuscript and preparing it for the printer.

Thanks are also due to S. Kalpakam, G. Rajamannar, K. S. Ramesh, and A. Rangan who have helped me in the correction of proofs and in the preparation of the index.

Finally, I am grateful to the Indian Institute of Technology, Madras for the facilities afforded for completing the work.

Madras   S. K. Srinivasan
February 1968

ACKNOWLEDGMENTS

The author is grateful to the Organo della Società Italiana di Fisica, Bologna, for permission to reproduce Figures 1 and 2 from *Nuovo Cimento*, **9,** 81 (1958).

The author also wishes to acknowledge his gratitude to the American Institute of Physics, New York, for permission to reproduce Figures 1 and 2 from the translated article by G. A. Timofeev in *Soviet Physics—JETP*, **14,** 1064 (1962).

CONTENTS

CHAPTER FOUR
Multiple Product Densities and Sequent Correlations

CHAPTER FIVE
Electromagnetic Cascades—Mathematical Techniques

CHAPTER SIX
Electromagnetic Cascades—Analytic and Computational Methods

CHAPTER SEVEN
Extensive Air Showers

CHAPTER EIGHT
Polarization in Cascades

CHAPTER NINE
Population Growth

Chapter 1

BRANCHING PHENOMENA

1. Introduction

The theory of stochastic processes is now finding increasing applications to problems in physics, astronomy, and biology which deal with processes involving some random element in their structure. Of course, the theory is an extension and enlargement of the scope of one of the old and well-established branches of mathematics—probability—and can therefore be considered as part of measure theory. While this approach is helpful in reaching a number of conclusions from analogous theorems in measure theory, occasionally it may be necessary to seek new methods and techniques for the formulation and solution of problems occurring in the realm of the natural sciences. The need for new methods and techniques has been very keenly felt in recent years, since the tendency to apply the results of stochastic theory to physical problems has increased considerably during the past decade. With the advent of branching phenomena and other allied techniques, the development of stochastic theory has received further impetus and a special branch of stochastic theory which has come to be known as "point processes" has come to the fore. The object of the present work is to present an account of some of these recent developments of stochastic processes that have had and continue to have considerable impact on cascade processes. We have taken the viewpoint of an applied mathematician and have not hesitated to use phenomenological methods in preference to abstract theory whenever it is advantageous to do so. Since the prime motivation for the development of the theory of stochastic point processes has stemmed from actual physical problems, we have found it worthwhile to discuss them in detail so that some of the special features of any particular problem may be (hopefully) conceived at a later date as part of the general pattern of the theory of point processes.

In the present chapter, which is of an introductory nature, we present an account of the general classification of stochastic processes and the classical method of dealing with probability frequency functions and other general properties of stochastic systems. We then discuss certain special features that arise from stochastic processes in which elements branch off insofar as their evolution is concerned. Some simple models of branching phenomena are then discussed with special reference to multiplicative processes. The final section deals with a brief introduction to multiplicative systems that are characterized by a continuous parameter.

2. Stochastic Processes—General

In ordinary probability theory, we are usually interested in a random variable x or an aggregate of random variables x_1, x_2, . . . and their corresponding distribution functions. Let us consider for convenience the case of a single random variable. If we imagine the distribution function to be a function also of a parameter t which we consider as one-dimensional for simplicity, it is natural to study the variation of the distribution function with t. The variation of t introduces a dynamical element into our problem and hence the "behavior" of the random variable for various values of t can be described as some process with some random element in its structure. A stochastic process* can be represented by the entire family of random variables $x(t)$ as t varies. The family can be considered as an ordered set if t is one-dimensional. Thus it involves merely the extension of the notion of an aggregate of random variables to the case when an aggregate is obtained by varying t. No difficulty is experienced in defining a process even when at a particular t, we have an aggregate $x_1(t)$, $x_2(t)$, . . . because t can be varied in each of these. However the interesting feature of $x(t)$ as distinguished from any aggregate, lies in the fact that it is possible to represent many physical processes involving some random element by establishing correspondence between $x(t)$ or a suitable aggregate $x_1(t)$, $x_2(t)$, . . . and such processes. In particular, when t is one-dimensional say for example, the time parameter, the "evolution" of the process can be represented by the sequence of random variables $x(t)$ obtained by varying t.

For our purpose, we can broadly divide stochastic processes into four classes taking the case where we have one random variable at any particular t:

(i) both x and t discrete

(ii) x discrete and t continuous

(iii) x continuous and t discrete

(iv) both x and t continuous.

* For an illuminating account, the reader is referred to a semi-expository discussion by Doob [1].

We can define

(i) $\pi(x_i; t_j)$ as the probability that $x(t_j) = x_i$, assuming $x_1, x_2, \ldots$ and $t_1, t_2, \ldots$ are the permissible values taken by x and t respectively,

(ii) $\pi(x_i; t)$ as the probability that $x(t) = x_i$,

(iii) $\pi(x; t_j)\, dx$ as the probability that $x(t_j)$ takes a value between x and $x + dx$ and

(iv) $\pi(x; t)\, dx$ as the probability that $x(t)$ has a value between x and $x + dx$.

The above class of processes can be further divided into two classes: Markovian processes and non-Markovian processes consisting of the residuary class. A process is called Markovian, if its development between t and $t + \Delta$ depends only upon the state of the system at t and independent of the "state" prior to t. Confining our attention to the probability frequency function (p.f.f.) we can express the p.f.f. at $t + \Delta$ purely in terms of the p.f.f. at t. We shall illustrate this point presently by displaying the classical Kolmogorov differential equation satisfied by the p.f.f. If we find a process to be non-Markovian, in many cases we can define a new process involving some additional elements which render the process Markovian. In fact, from a physical point of view, it is eminently reasonable to visualize a non-Markovian process as the projection of a Markovian process, the non-Markovian nature stemming from the absence of the extra "coordinates" lost by projection.

To obtain the differential equation satisfied by the p.f.f., we analyze the outcome of different possibilities in the interval $(t, t + \Delta)$. In (ii) above, let $\pi(x_i | x_k, t)$ denote the probability that $x(t) = x_i$ given $x(0) = x_k$. We shall assume that (see, for example, Feller [2])

$$(2.1) \qquad \lim_{\Delta \to 0} \pi(x_i | x_k; \Delta) = R_{ik}\Delta + o(\Delta)$$

so that the probability that the random variable continues to take the value x_i during Δ is $1 - \sum_{i \neq k} R_{ik}\Delta + o(\Delta)$. If we further assume that the process is homogeneous with respect to t, then we have

$$(2.2) \qquad \pi(x_i | x_k; t + \Delta) = (1 - \sum_{j \neq i} R_{ji}\Delta)\pi(x_i | x_k; t)$$

$$+ \sum_{j \neq i} \pi(x_j | x_k; t) R_{ij}\Delta$$

$$+ o(\Delta)$$

Proceeding to the limit as Δ tends to zero, we obtain the celebrated Kolomo-gorov forward differential equation

$$(2.3) \qquad \frac{\partial \pi(x_i | x_k; t)}{\partial t} = - \sum_{j \neq i} R_{ji} \pi(x_i | x_k; t)$$
$$+ \sum_{j \neq i} \pi(x_j | x_k; t) R_{ij}$$

describing the change in the p.f.f. of the random variable $x(t)$ as t increases.

There is yet another approach which consists in analyzing the outcome of various possibilities in the interval $(0, \Delta)$ and expressing the p.f.f. at t in terms of the p.f.f. at $t - \Delta$. Thus we have

$$(2.4) \qquad \pi(x_i | x_k; t) = (1 - \sum_{j \neq k} R_{jk} \Delta) \pi(x_i | x_k; t - \Delta)$$
$$+ \sum_{j \neq k} \pi(x_i | x_j; t - \Delta) R_{jk} \Delta + o(\Delta)$$

where we have made tacit use of the homogeneous nature of the process. Proceeding to the limit as Δ tends to zero, we obtain the backward differential equation

$$(2.5) \qquad \frac{\partial \pi(x_i | x_k; t)}{\partial t} = -\pi(x_i | x_k; t) \sum_{j \neq k} R_{jk} + \sum_{j \neq k} \pi(x_i | x_j; t) R_{jk}.$$

If x is capable of assuming continuous values, (2.3) and (2.5) are still valid provided we replace the sums by appropriate integrals. In the case when the transition probabilities are t-dependent, equation (2.4) is still applicable with obvious modifications. However, in order to obtain the backward equation similar to (2.5), we should apply (2.4) successively over all possible Δ's and this will lead to an integral equation. We will have occasion to deal with such a method in Chapter 5.

3. Multiplicative Processes

Let us assume that "events" occur along the t-axis (t being one-dimensional) so that the state of the system at t is characterized by the number of events that have occurred up to t. If the probability of an event occurring between t and $t + dt$ is a function of the number of events that have occurred between 0 and t, the process is said to be multiplicative. In a general setting, an event can be identified to be the "birth" of an "entity" and the state of the system characterized by the number of entities that exist at t. Then the total number of events between 0 and t is not the same as the number of entities that exist at t if at any t there is a non-zero probability of absorption or death of an entity. The process becomes non-Markovian in character if we assume the probability of death to depend on the life-span of the individual entity. We shall deal with such population processes in the final chapter of this monograph. In this section, we shall deal with two simple types of multiplicative processes which are Markovian.

3.1. POISSON PROCESS

The simplest example of a multiplicative process is obtained by assuming that events occur along t-axis, the probability of an event happening between t and $t + dt$ is λdt.* The parameter λ is independent of the state of the system at t which is characterized by the number of events that have occurred between 0 and t. If λ is independent of t, the process is called a simple Poisson process or homogeneous Poisson process. If however λ is a function of t, then it is called an inhomogeneous Poisson process. If $\pi(n, t)$ is the probability that n events occur between 0 and t, we can use the Markovian nature of the process to obtain

$$(3.1) \qquad \frac{\partial \pi(n, t)}{\partial t} = -\lambda(t)\pi(n, t) + \lambda(t)\pi(n - 1, t).$$

Explicit solution of $\pi(n, t)$ is obtained by solving (3.1) recursively. Alternatively, we can use the method of generating function by defining

$$(3.2) \qquad G(u, t) = \sum_n u^n \pi(n, t)$$

so that $G(u, t)$ satisfies the equation

$$(3.3) \qquad \frac{\partial G(u, t)}{\partial t} = -\lambda(t)G(u, t) + \lambda(t)uG(u, t)$$

whose solution is given by

$$G(u, t) = G(u, 0) \exp \left[(u - 1) \int_0^t \lambda(t')\, dt' \right].$$

* There are a number of equivalent methods of characterizing a Poisson process. For an excellent account of the same, the reader is referred to the monograph of Parzen [3].

Since we have defined $\pi(n, t)$ as the p.f.f. of the number of events that have occurred between 0 and t, we have $G(u, 0) = 1$ corresponding to the initial condition $\pi(n, 0) = \delta_{n0}$.
Thus we obtain

$$(3.4) \qquad \pi(n, t) = e^{-\Lambda(t)}[\Lambda(t)]^n/n!$$

where

$$(3.5) \qquad \Lambda(t) = \int_0^t \lambda(t')\, dt'.$$

Examples of Poisson process can be found in all realms of natural phenomena. The emissions of radiations from a radioactive source constitute events the counting process of which is a Poisson process. If we consider simple types of electronic valves, there do exist current fluctuations usually called noise. This is due to the emission of electrons from the heated cathode, the number $n(t)$ of electrons emitted in a time interval $(0, t)$ constituting a Poisson process in which the average rate of emission of electrons remains a constant. Another example of a Poisson process is provided by the machine failures in a system consisting of a number of components each of which has a certain lifetime. The lifetimes, in general, are independent and obey an exponential distribution. In such cases, the number of failures or breakdowns of the system constitute a Poisson process. Again in ecology, the distributions in space of plants are describable by Poisson processes. We shall show in Chapter 4 how the Poisson nature of the distribution of molecules in velocity space leads to physically meaningful types of particle statistics which, in turn, lead to the celebrated Fermi and Bose spectra for the energy distribution.

3.2. YULE-FURRY PROCESS

In an attempt to estimate the fluctuation in the number of particles produced in a shower of electrons passing through lead, Furry [4] proposed and solved the following problem:

The probability that in traversing thickness dt one particle is converted into two is just dt. If one particle enters a sheet of thickness t, what is the probability $P(n, t)$ that n particles will emerge?

This is another example of a multiplicative process in which the probability of an event occurring between t and $t + dt$ is $n\, dt$ where n is the number of events that have occurred between 0 and t. The process is named after Furry who has made pioneering contributions in the direction of estimating the size fluctuation of electron-photon showers. The process is of considerable importance in the study of population growth since the production of an extra particle can also be considered as binary fission. In fact, Yule [5] has studied this problem in connection with the evolutionary theory of population growth. Although he had obtained explicit expression, as early as 1924, for the probability frequency function governing the number of individuals at any time t produced by a single ancestor present at $t = 0$, its relevance to the fluctuation problem of cosmic rays was not noticed until Furry re-derived the results in the language of cascade theory.

If we use the forward differential equation method, we obtain

$$(3.6) \qquad \frac{\partial P(n, t)}{\partial t} = -nP(n, t) + (n - 1)P(n - 1, t), n > 1$$

$$\frac{\partial P(1, t)}{\partial t} = -P(1, t)$$

where we have assumed that at $t = 0$ there is one particle. Defining $g(u, t)$ as the generating function of $P(n, t)$ we find

$$(3.7) \qquad \frac{\partial g(u, t)}{\partial t} = u(u - 1)\frac{\partial g(u, t)}{\partial u}$$

with the initial condition

$$(3.8) \qquad g(u, 0) = u.$$

Solving the partial differential equation (3.7), we obtain

$$(3.9) \qquad g(u, t) = u\,e^{-t}/[1 - u(1 - e^{-t})]$$

from which we find

$$(3.10) \qquad P(n, t) = e^{-t}(1 - e^{-t})^{n-1}, \qquad n > 1.$$

In the present case, we can also use the method of backward differential equation*. To achieve this, we observe that in the infinitesimal interval $(0, \Delta)$, the particle present at $t = 0$ undergoes fission with probability Δ in which case the two fission products act as independent primaries and produce n particles in all the interval (Δ, t). Thus using the homogeneous nature of the process, we find

$$(3.11) \qquad P(n, t) = (1 - \Delta)P(n, t - \Delta)$$
$$+ \sum_{n_1 + n_2 = n} \Delta P(n_1, t - \Delta)P(n_2, t - \Delta) + o(\Delta).$$

* This was noticed by Ramakrishnan [6].

Proceeding to the limit as $\Delta \rightarrow 0$, we find

$$(3.12) \qquad \frac{\partial P(n, t)}{\partial t} = -P(n, t) + \sum_{n_1 + n_2 = n} P(n_1, t)\, P(n_2, t)$$

so that $g(u, t)$ satisfies the equation

$$(3.13) \qquad \frac{\partial g(u, t)}{\partial t} = -g(u, t) + [g(u, t)]^2$$

which can be shown to lead to the solution (3.9). In a more general situation, the backward differential equation is not capable of explicit solution in view of its nonlinear structure and hence has not proved to be very popular. This point will be discussed in detail in Chapter 6.

We can consider an inhomogeneous Furry process by assuming that the probability of fission of a particle during the infinitesimal interval $(t, t + dt)$ is $\lambda(t)\, dt$. In this case, while it may be difficult to find an equation analogous to (3.13), it is quite easy to obtain the forward differential equation. In fact (3.6) and (3.7) are valid in the present case, provided we introduce an extra factor $\lambda(t)$ in the right-hand side. The solution for $g(u, t)$ is given by

$$(3.14) \qquad g(u, t) = u\, e^{-\Lambda(t)}/[1 - u(1 - e^{-\Lambda(t)})].$$

The Furry process can be generalized so as to make the probability of production of a particle at a time t to depend on the number as well. This leads us to the general birth process. If we assume that $\lambda_n(t)\, dt + o(dt)$ is the probability that a particle is produced between t and $t + dt$, n being the population size at t, $P(n, t)$ satisfies the equation

$$(3.15) \qquad \frac{\partial P(n, t)}{\partial t} = -\lambda_n(t)\, P(n, t) + \lambda_{n-1}(t)\, P(n - 1, t), \qquad n > 0$$

$$\frac{\partial P(0, t)}{\partial t} = -\lambda_0(t)\, P(0, t).$$

If $\lambda_n(t)$ is independent of t, the solution of $P(n, t)$ can be obtained explicitly (see, for example, Bartlett [7]).The necessary and sufficient conditions for equations (3.15) to have a unique and valid solution has been investigated by Feller [2]. When we introduce the probability of death, the problem becomes a little more complicated, and we shall discuss this problem in Chapter 9 in detail.

3.3. AGE DEPENDENT PROCESSES

So far we have taken into account the inhomogeneous nature of the process by assuming that the transition probabilities are dependent on t. There is yet another type of inhomogeneity arising from the dependence of transition probabilities on the age of the different individuals constituting the population. Thus the process becomes non-Markovian in character and a more complete description of the state of the system is necessary to render it Markovian. In this chapter, which is introductory in nature, we shall show how in a simple model we can use the backward differential equation method to obtain the probability frequency function of the size of the population at any time t.

Let us assume that at $t = 0$ there is a primary individual of age x_0 and that any individual of age x at time t produces another between t and $t + \Delta$ with probability $\lambda(x)\Delta + o(\Delta)$. If $\pi(n, x_0, t)$ is the probability that n individuals are found at t, we observe that the process is still homogeneous with respect to t and the age dependence is reflected by the fact that π is a *functional of λ*. We bring out the functional dependence of π on λ by observing that in the infinitesimal interval $(0, \Delta)$ there is a non-zero probability that the primary produces a secondary in which case we have two independent primaries at Δ. Thus we have two independent population processes of duration $t - \Delta$ due to the individuals of age 0 and $x_0 + \Delta$. Using arguments similar to the deduction (2.5), we obtain

$$(3.16) \qquad \frac{\partial \pi(n, x_0, t)}{\partial t} = -\lambda(x_0)\pi(n, x_0, t) + \frac{\partial \pi(n, x_0, t)}{\partial x_0}$$

$$+ \lambda(x_0) \sum_{n_1 + n_2 = n} \pi(n_1, x_0, t)\, \pi(n_2, 0, t).$$

The nonlinear nature of the above equation renders its explicit solution difficult. However it can be used to obtain information regarding the first few moments and we shall return to this problem in Chapter 5. The integral form of (3.16) was first observed by Bellman and Harris [8] and has been developed to a high degree of sophistication (see, for example, Bellman *et al.* [9]) and is well known under the name invariant imbedding technique. As we will have many occasions to resort to this technique in this monograph, we shall deal with this in detail in Chapter 5.

3.4. CONTINUOUS PARAMETRIC SYSTEMS

In the introductory part of this section, we have characterized a multiplicative process as one in which the transition probability is a function of the number of events that have occurred between 0 and t. There is an alternative but useful mode of description of the same process. We can interpret the process as a distribution of random points on a line by identifying every random point with an event. Such processes have come to be known as *point processes*. The process can be visualized to be a continuous parametric system, the continuous parameter being t in this case. In the present case, the process evolves with respect to t. If we consider the age dependent evolution of population growth as outlined above, the continuous parameter will signify the age so that we deal with a population whose individuals are distributed over x-space, the process itself progressing or evolving with respect to t. In such a case, there arises a difficulty since only probability densities can be attached to particular values of x and not non-zero probabilities in the sense that probability measures cannot be assigned to the Borel subsets of x-space. While the densities attached to particular values of x are useful in obtaining the mean value of the numbers of the individuals distributed in prescribed ranges of x, they do not enable us to estimate the fluctuation about the mean which is an important criterion. This has been the starting point of the theory of point processes and a number of techniques mathematical and phenomenological in nature have been developed during the past two decades. The various methods that are available for handling general point processes will be discussed in detail in the next two chapters.

REFERENCES

1. J. L. Doob, *Amer. Math. Monthly*, **49** (1942), 648.
2. W. Feller, *An Introduction to Probability Theory and Its Applications*, John Wiley, New York, 1950.
3. E. Parzen, *Stochastic Processes*, Holden-Day Inc., San Francisco, 1962.
4. W. H. Furry, *Phys. Rev.*, **52** (1937), 569.
5. G. U. Yule, *Phil. Trans. Roy. Soc.* (London), **B 213** (1924), 21.
6. A. Ramakrishnan, *Probability and Stochastic Processes*, (*Handbuch der Physik*, Vol. III), Springer, Berlin, 1959.
7. M. S. Bartlett, *Stochastic Processes*, Cambridge, 1955.
8. R. E. Bellman and T. E. Harris, *Proc. Nat. Acad. Sci. U.S.A.*, **34** (1948), 601.
9. R. E. Bellman, R. Kalaba, and G. M. Wing, *J. Math. Phys.*, **1** (1960), 280.

Chapter 2

POINT PROCESSES—GENERAL APPROACH

1. Historical Remarks

In probability theory we usually study a random variable or a sequence of random variables and their corresponding distribution functions. Such studies of random variables and their distributions have been extremely useful in many classical situations such as the random walk, gambler's ruin, and the extinction of peers (see, for example Feller [1]). However, of late, new situations have arisen in physical, biological, and social sciences wherein we deal with a random variable representing the number of entities distributed over a continuous infinity of states characterized by a parameter x, the random variable itself "evolving" or progressing with respect to another parameter t which may be continuous or discrete. If we denote by $N(x, t)$ the random variable representing the number of entities with parametric values less than or equal to x corresponding to the parametric value t, then a study of the basic process consists in obtaining information regarding the probability frequency function (p.f.f.) of $N(x, t)$ at one or more values of t and perhaps the simultaneous p.f.f. of $N(x, t)$ corresponding to several values of t. Such distributions were first studied by Kendall [2] in connection with the fluctuation of the size of populations. Since $N(x, t)$ is a function of points in t-space and intervals of x-axis, such processes are called point processes (see Bartlett [3], Moyal [4]). The theory of point processes was first studied by Wold [5] in connection with the properties of an ordered sequence of points randomly located on a line and was subsequently studied by Cox [6], Smith [7], McFadden [8], and more recently by Beutler and Leneman [9]. However, its wide applicability in a variety of phenomena has been reported by Bartlett [3, 10, 11] and Moyal [4, 12]. In fact, the techniques of point processes were developed in the late thirties and early forties of this century in an entirely different context by Yvon (see de Boer [13], Nordsieck, Lamb, and Uhlenbeck [14], Bogoliubov [15], Kirkwood [16], and Born and Green [17]. All these authors with the exception of Nordsieck, Lamb, and Uhlenbeck have been mainly concerned with the formulation of the kinetic theory of fluids while

Nordsieck, Lamb, and Uhlenbeck have dealt with the fluctuation of the number of particles in an electromagnetic cascade through certain cluster and correlation functions introduced by an analogy with the discrete parametric space. These correlation functions have been systematized and rationalized by Bhabha [18] and Ramakrishnan [19] who formulated the product density approach to stochastic point processes in general. The product density technique has been generalized by Moyal [12] to cover the case where the x-space is no longer Euclidean. The simplicity of the product density technique has led to a different type of generalization in sequent densities [20, 21, 22] which are defined for a sequence of points in t-space, the parameter with respect to which the underlying stochastic process evolves. In their contributions of recent years Bartlett [11] has discussed the role of the point processes in renewal theory and Kendall [23] has observed that these techniques will have a big impact on queuing theory and other associated problems. We propose to deal with these numerous developments in point processes in this chapter. Since the product densities enjoy a uniquely privileged position in applications particularly in view of their possession of direct probabilistic interpretation, we shall deal with them in a detailed manner in the next chapter.

The plan of the present chapter is as follows. We shall define the point distribution and the density functions associated with them in Section 2. Section 3 contains an account of Bhabha's theory of point distribution leading to the product density techniques. Finally, we present the method of characteristic functional and outline a method of extracting the statistical properties of the point processes from such a characteristic functional.

2. Point Distribution and Janossy Densities

As mentioned in Section 1, point processes are stochastic processes specified in relation to objects or entities each labeled with the random value of a continuous parameter x. The process itself may progress or evolve with respect to another parameter t as assumed in the introductory remarks; however, this is not necessary for our present purposes since in the case of stationary processes, the (past) evolution is of no significance. It will be assumed that x is a point in p-dimensional Euclidean space R. In many cases, x will be restricted to some part of R which we shall denote by X. Thus if x is the age of an individual in a population or the energy of a cosmic ray particle, R is the real line and X is the non-negative part of R.* The immediate difficulty is

---

* Of course, X will be a member of the Borel subsets of R.

that only probability densities can be attached to particular values of the parameter x and not non-zero probabilities. It is precisely on account of such a situation that Bartlett ([3], see page 78) prefers to call them point processes. The difficulty is eliminated automatically if we restrict x to a finite set of distinct points $x_1, x_2, \ldots, x_m$ in X. For then we can associate a positive integer with each x_i and it is possible to define non-zero probabilities for such point distributions. In this manner, one can go ahead and build up a theory with suitable techniques capable of adaptation to any particular problem. Moyal [12] attempted to develop such a method by constructing a suitable measure space out of the Borel subsets of R. In fact, he considered a more general situation when R can be an abstract space. Thus Moyal's attempts in such directions place some of the techniques that have already been well developed and used in specific physical problems by Janossy [24], Ramakrishnan [19], and Srinivasan [25] on a firm mathematical footing.

As discussed earlier, the difficult nature of the point process was first realized in the stochastic problem of population growth and to a much greater extent in the fluctuation problem of cosmic ray cascades. An outstanding contribution in the direction of resolution of the difficulties of a pathological nature is that of Janossy [24] whose work is primarily responsible for unraveling various features relating to the development of electromagnetic and nucleon cascades. In fact, more than fifty papers that have been published by Messel and his collaborators (see Messel [26]) on the longitudinal and lateral development of showers are based on the density function technique of Janossy. If t is the parameter with respect to which the stochastic process evolves, t standing for thickness of matter traversed by a cosmic ray particle (since the problem is originally treated as one-dimensional), the lateral spread of the shower being negligible at high energies, and x is the parameter space (energy) over which the particles are distributed, we can define the Janossy density of order n by $J_n(x_1, x_2, \ldots, x_n; t)$ where $J_n(x_1, x_2, \ldots, x_n; t)$ $dx_1 \, dx_2 \cdots dx_n$ denotes the probability that there are exactly n particles distributed in the parametric space with parametric values between x_i and $x_i + dx_i (i = 1, 2, \ldots n)$. The particles described here are indistinguishable except for the parametric values assumed by them and hence J_n has the property

$$(2.1) \qquad \frac{1}{n!} \int dx_1 \int dx_2 \cdots \int dx_n \, J_n(x_1, x_2, \ldots, x_n; t) = P(n)$$

where $P(n)$ is the probability of finding n particles distributed over the entire x-space. Thus while it is convenient to deal with $J_n/n!$ it is J_n that has physical significance since the particles are indistinguishable.

2.1. A SIMPLE EXAMPLE

The Janossy function will also enable us to calculate $P(n; a, b)$ the probability of finding n particles whose parametric values lie in a prescribed interval $[a, b]$. This is obtained by integrating J_n in the appropriate region which is nothing but the n-fold Cartesian product of the intervals $[a, b]$. To illustrate the method, we shall deal with the nucleon cascade in which there is only one type of particle (nucleon). When a proton of the primary cosmic radiation strikes a nucleus, it knocks one or more nucleons as well as mesons which are assumed to have no part in the production of further nucleons. The new nucleons then produce further nucleons by a similar process and so on.

Given the initial energy spectrum of nucleons at $t = 0$ (t represents the thickness of matter traversed), and that a nucleon of energy E drops to an energy lying between E_i and $E_i + dE_i$, creating a recoil nucleon of energy between E_j and $E_j + dE_j$ with a probability per unit distance given by $\omega(E; E_i, E_j)\, dE_i\, dE_j$ our object is to calculate the probability distribution of the number of nucleons at t above a certain energy. Since we cannot distinguish between recoil and incident nucleons we can define

$$(2.2) \qquad W(E; E_i, E_j) = \omega(E; E_i, E_j) + \omega(E; E_j, E_i)$$

where $W(E; E_i, E_j)\, dE_i\, dE_j$ represents the probability per unit distance that a particle of energy E is replaced by a pair, one of which has an energy E_i and another E_j. Using Janossy's functions we can write the diffusion equation for the stochastic process as

$$(2.3) \qquad \frac{\partial J_m(E_1, E_2, \ldots, E_m; t)}{\partial t} = -J_m(E_1, E_2, \ldots, E_m; t) \sum_{i=1}^{m} W'(E_i)$$

$$+ \sum_{i=1}^{m} \sum_{\substack{j=1 \\ i \neq j}}^{m} \int_{E} J_{m-1}(E_1'', E_2'', \ldots, E_{m-2}'', E; t)\, W(E; E_i, E_j)\, dE$$

where

$$W'(E) = \int_{E_i} \int_{E_j} W(E; E_i, E_j)\, dE_i\, dE_j$$

and $J_{m-1}(E_1'', E_2'', \ldots, E_{m-2}'', E; t)$ is a Janossy function of order $(m - 1)$; E_k'' for $k = 1, 2, \ldots, m - 2$ is supposed to represent the values of E_k, $k = 1, 2, \ldots, m$ omitting E_i and E_j. This is slightly different from the notation adopted by Janossy, who writes E_k' instead of E_k.

Assuming that $W(E; E_i, E_j) \, dE_i \, dE_j$ can be expressed in the form $W(\mathscr{E}_i, \mathscr{E}_j) \, d\mathscr{E}_i \, d\mathscr{E}_j$ where $\mathscr{E}_i = E_i/E$, $\mathscr{E}_j = E_j/E$ Janossy solved his equation (2.3) using the Mellin transform technique.

It might appear from the above that once J_n is obtained for all n in any process, all the physically relevant information about the process can be obtained. However, the explicit solution for J_n for any n is a difficult task in a general situation, though it might be possible to calculate for the first few n starting from $n = 1$. In view of this difficulty, the problem of finding even the first few moments of the number distribution in a more realistic problem like the electromagnetic cascade is extremely difficult. In fact, Janossy himself abandoned this method and resorted to what is now known as G-function technique—a topic that will be dealt with in Chapter 5. In this respect, the product density technique of Ramakrishnan is decisively superior since it connects the r-th moment of the number distribution in terms of product densities of degree $<r$.

Distribution functions similar to Janossy densities have been independently used by Bogoliubov [15] in his treatment of problems of dynamical theory in statistical physics. Starting from Janossy densities of molecules, Bogoliubov uses a limiting procedure to obtain the analogue of the Boltzmann equation. Born and Green and Kirkwood have proposed a kinetic theory of liquids using similar distribution functions. It appears that many of these features had been anticipated by Yvon as early as 1937 (see reference [13]).

3. Bhabha's Theory of Point Distribution

As we mentioned in the introductory remarks, Bhabha, who was primarily motivated by the cosmic ray problem, first dealt with the theory of continuous parametric stochastic systems by proposing a discrete model having the properties of a continuum and later, by generalizing the concept, introduced the point distribution. The same method was simultaneously proposed by Janossy who did not pay attention to the calculation of moments of the number distribution. Later on, the connection between the two approaches was made clear by Ramakrishnan [27] who obtained an explicit expression for the product densities in terms of the Janossy densities. Scott [28] claims to have had similar ideas as early as 1946; his results similar to Janossy densities and their relation to the general moment problem were published in 1951. In this section, we shall present an account of Bhabha's method since it is complementary to the method of Janossy himself to the problem of determining the distribution of the number of particles in a given energy interval.

3.1. A DISCRETE MODEL HAVING THE PROPERTIES OF A CONTINUUM OF STOCHASTIC VARIABLES

We present here some of the first results in the quest for a useful theory of point distribution. The first attempts in this direction were made by Scott and Uhlenbeck [29] who demonstrated a method of obtaining the mean square number of entities distributed over a continuous parametric space. Bhabha and Ramakrishnan (see reference [30]) systematized the approach by building up a system of particles occupying a discrete number of states (finite or denumerably infinite) which presents features analogous to the continuous system. They have assumed that in every state there can be 1 or 0 particles. In other words, we have a sequence of stochastic variables n_1, n_2, ..., n_i, .. each of which can assume the values 0 or 1. In such a case, we can define the probability that n_i assumes the value 1 as $_{(1)}P_i$. We can define a joint probability of degree two that n_i assumes the value 1 and n_j assumes the value 1 as $_{(2)}P_{ij}$. If ν is defined by

$$(3.1) \qquad\qquad \nu = \Sigma n_i$$

then the r-th moment of ν is given by

$$(3.2) \qquad \overline{\nu^r} = \Sigma \{ C_s^r \sum_i \sum_j \cdots \Sigma_{(s)} P_{ijk} \ldots \}.$$

In particular, we have

$$(3.3) \qquad \overline{\nu} = \Sigma_{(1)} P_i, \quad \overline{\nu^2} = \Sigma_{(1)} P_i + \sum_{\substack{i \ j \\ i \neq j}} \Sigma_{(2)} P_{ij}$$

so that it can be concluded that if these stochastic variables tend to infinity keeping $\overline{\nu}$ finite, then ν approaches a Poisson distribution. Koopman [31] has considered such a sequence of variables possessing Markovian properties. The sequence becomes Markovian if the probability that n_j assumes the value 0 or 1 given that n_i has assumed the value 0 or 1 ($i < j$) does not depend on the additional information about the value n_k has assumed ($k < i$). Accordingly, if we define

$$a_J = \text{prob} \ (n_J = 1 \mid n_{J-1} = 1)$$

(3.4)

$$b_J = \text{prob} \ (n_J = 1 \mid n_{J-1} = 0)$$

we can obtain explicit expressions for the probabilities mentioned above in terms of a_J and b_J. In fact, Koopman has obtained the probability generating function of the random variable ν by assuming that a_J and b_J are independent of J.

To deal with the system whose states are labeled by a continuous parameter x, we divide the domain of x into a large number, say ν, of small segments of equal length δ extending respectively from 0 to x_1, x_1 to x_2, x_2 to x_3, and so on. If n_i denotes the actual number of particles in the i-th segment, then the state of the assembly is completely characterized by giving a set of integral values to the complete set of ν numbers $n_1, n_2, \ldots, n_\nu$. With every such state $(n_1, n_2, \ldots, n_\nu)$ we associate a real number $\pi(n_1, n_2, \ldots, n_\nu)$ giving the probability of the assembly in that state, and a knowledge of all such probabilities will enable us to determine all the statistical properties of the system. For example, we can define the mean of the product $[n_i]^r[n_j]^s \cdots$ through the equation

$$(3.5) \qquad \overline{[n_i]^r[n_j]^s \cdots} = \Sigma [n_i]^r[n_j]^s \cdots \pi(n_1, n_2, \ldots, n_\nu)$$

where the summation is over all integral values of $n_i, n_j, \ldots$.

If $N_{ij} = \sum_{k \neq i}^{j} n_k$, then it follows

$$(3.6a) \qquad \overline{N_{ij}} = \sum_{k=i}^{j} \overline{n_k}.$$

$$(3.6b) \qquad \overline{N_{ij}^2} = \sum_{k=i}^{j} \overline{n_i^2} + \sum_{k=i}^{j} \sum_{\substack{l=i \\ k \neq l}}^{j} \overline{n_k n_l}.$$

It is interesting to observe that while it is convenient to assume that $\overline{n_i^2} = \overline{n_i}$ in the limit as δ tends to zero, it need not be so in general. Assuming the variation of $\overline{n_k}$, $\overline{n_k^2}$, $\overline{n_k n_l}$ from one state to the next tends to zero as δ tends to zero, we can replace the summations in (3.6) by integrations so that

$$(3.7a) \qquad \overline{N(x_i, x_j)} = \int_{x_i}^{x_j} \overline{n(x)}\, dx$$

$$(3.7b) \qquad \overline{N(x_i, x_j)^2} = \int_{x_i}^{x_j} \overline{n^2(x)}\, dx + \int_{x_i}^{x_j}\int_{x_i}^{x_j} \overline{n_2(x, x')}\, dx\, dx'$$

where

$$(3.7c) \qquad \bar{n}(x_i) = \lim_{\delta \to 0} \frac{\overline{n_i}}{\delta}, \quad \overline{n^2(x_i)} = \lim \frac{\overline{n_i^2}}{\delta}$$

$$\overline{n_2(x_i, x_j)} = \lim \overline{n_i n_j}/\delta^2$$

provided these limits do exist. Since $\overline{n_i^2} \neq \overline{n_i}$ even in the limit as $\delta \to 0$ in general $\overline{n^2(x)} \neq \overline{n(x)}$. Scott and Uhlenbeck [29] had assumed it to be so, and in this sense they anticipated the appropriate description of the cascade theory. As Bhabha [18] observes "The crux of the matter is that the continuous parametric assembly which is arrived at as the limit of a discrete state assembly in the above way has far more general properties than most of the actual assemblies with which one has to deal in nature, and in consequence, certain general properties which the latter assemblies possess are absent in the more general assemblies described above." In fact, when we introduce multiple points or have to deal with associated processes it may be necessary to relax the condition $\overline{n_i^2} = \overline{n_i}$. This was noted by Kendall [2] in his study of population models and we shall return to this again in the next section when we will discuss the method of characteristic functionals. In the next subsection, we shall present Bhabha's method of approach to arrive at the product density description of a class of continuous parametric systems.

3.2. A SPECIAL MODEL OF A CONTINUOUS PARAMETRIC SYSTEM

The starting point of Bhabha is the recognition of the fact that the possible eigen state of the continuous parametric assembly is one in which there are k systems present with the parametric values in the intervals $(x_1, x_1 + dx_1)$, $(x_2, x_2 + dx_2), \ldots, (x_k, x_k + dx_k)$, $(x_1 < x_2 < \cdots < x_k)$. Accordingly, he defined the function $\pi(x_1, x_2, \ldots, x_k)$ where $\pi(x_1, x_2, \ldots, x_k)\, dx_1\, dx_2 \cdots dx_k$ is interpreted to be the probability of the particular state of the assembly; π satisfies the condition

$$(3.8) \qquad \pi_0 + \sum_{k=1}^{\infty} \int_0^{\infty} dx_k \int_0^{x_k} dx_{k-1} \cdots \int_0^{x_2} dx_1\, \pi(x_1, x_2, \ldots, x_k) = 1$$

where π_0 stands for the probability of the eigen state in which no system is present. Let p be some property of the assembly whose value for the eigen state $(x_1, x_2, \ldots, x_k)$ is $p(x_1, x_2, \ldots, x_k)$. If X_k is the k-dimensional space of all eigen states defined above, then the average value of p over X_k is given by

$$(3.9) \quad \bar{p}(X_k) = \iint \cdots \int dx_1\, dx_2 \cdots dx_k\, \pi(x_1, x_2, \ldots, x_k)\, p(x_1, x_2, \ldots, x_k).$$

If we remove the restriction that $x_1, x_2, \ldots, x_k$ are ordered and denote by X'_k the extended space, then

(3.10)
$$\bar{p}(X'_k) = k!\bar{p}(X_k)$$

and (3.9) can be generalized as

(3.11)
$$\bar{p} = \pi_0 p_0 + \sum_1^{\infty} \frac{1}{k!} \int_0^{\infty} dk_k \int_0^{\infty} dx_{k-1} \cdots \int_0^{\infty} dx_1$$
$$p(x_1, x_2, \ldots, x_k)\, \pi(x_1, x_2, \ldots, x_k).$$

Next we introduce the sawtooth function T by

(3.12)
$$T(x|x_1|x') = \begin{cases} 0 & x_1 < x \\ 1 & x \leqslant x_1 \leqslant x' \\ 0 & x_1 > x' \end{cases}$$

so that T satisfies the condition

(3.13)
$$[T(x|x_1|x')]^2 = T(x|x_1|x').$$

If $N(x, x')$ denotes the number of systems of parametric values between x and x', then the number of systems in the eigen state $(x_1, x_2, \ldots, x_k)$ is given by

(3.14)
$$N(x|x_1, x_2, \ldots, x_k|x') = \sum_{i=1}^{k} T(x|x_i|x').$$

Then it follows that

(3.15)
$$[N(x|x_1, x_2, \ldots, x_k|x')]^2 = N(x|x_1, x_2, \ldots, x_k|x')$$
$$+ \sum_{\substack{i=1 \\ i \neq j}}^{k} \sum_{j=1}^{k} T(x|x_i|x')\, T(x|x_j|x').$$

We can obtain the mean values of $N(x, x')$ and $[N(x, x')]^2$ by substituting the right-hand side of (3.14) and (3.15) in (3.11). Thus we obtain

(3.16)
$$\overline{N(x, x')} = \int_x^{x'} dx''\, \bar{n}(x'')$$

where

$$\bar{n}(x) = \sum_{k=1}^{\infty} \frac{1}{(k-1)!} \int_0^{\infty} dx_{k-1} \cdots \int_0^{\infty} dx_1\, \pi(x_1, x_2, \ldots, x_{k-1}, x).$$

$\bar{n}(x)\, dx$ can be interpreted to be the mean number of systems whose para-metric values lie between x and $x + dx$. Likewise we obtain

$$(3.17) \qquad \overline{N(x, x')^2} = \overline{N(x, x')} + \int_x^{x'} dx'' \int_x^{x'} dx''' \, \overline{n_2(x'', x''')}$$

where $n_2(x'', x''')$ is defined by

$$(3.18) \quad n_2(x'', x''') = \sum_0^\infty \frac{1}{k!} \int_0^\infty dx_k \cdots \int_0^\infty dx_1 \, \pi(x_1, x_2, \ldots, x_k, x'', x''').$$

In these equations $n_2(x'', x''')$ can be interpreted as the density of the mean number of systems whose parametric values are disjoint and lie in the intervals $(x'', x'' + dx'')$ and $(x''', x''' + dx''')$. These relations have been generalized by Bhabha to higher moments of the variable $N(x, x')$ but he was not successful in calculating explicitly the degeneracies arising in the overlap of the variables x. Ramakrishnan has taken a slightly different approach by dealing with the random variable $N(x)$ denoting the number of systems each with a parametric value less than or equal to x. By studying the properties of $dN(x)$ he has been able to present a better treatment of the statistical problem and to arrive at an explicit expression for the general moment of $N(x, x')$, and this will be discussed in the next chapter. *In conclusion, we wish to remark that the function defined by Bhabha is nothing but the Janossy function (introduced in Section 2) which was perhaps well known to the kinetic theorists as early as 1946.**

4. The Method of Characteristic Functional

The idea of a characteristic functional has been put forward by Bochner [32] and Le Cam [33] in 1947. Bochner introduced an over-all characteristic function $\phi[h]$ taking numerical values and defined for all the elements of a certain vector space. Le Cam was concerned with a problem dealt with by Halphen (see reference [33]) connecting the river water flow to rainfall. He associated with each random variable $x(t)$ and a function $s(t)$ the real number defined by

$$(4.1) \qquad \Phi_x[s(t)] = E\,\{\exp i \int x(t)\, s(t)\, dt\}.$$

---

* In particular, the formula for the second-order moment appears to have been explicitly derived by Yvon (see reference [13]) as early as 1937.

Le Cam names (4.1) as "functionnelle caractéristique." A complete mathematical account of the characteristic functional was given later in the monograph of Bochner [34].

As we have observed in Section 2, the difficult nature of the continuous parametric systems arises from our *inability* to define a function $\pi(n_1, x_1; n_2, x_2; \cdots; n_m, x_m)$ denoting the probability that n_1 systems are characterized by the parameter x_1, and n_2 systems by the parameter $x_2 \ldots$. It is precisely this situation that enables us to define a functional by

$$(4.2) \qquad \Phi = E \{\exp i \int \theta(x) \, dN(x)\}$$

where $N(x)$ is the random variable denoting the number of systems (particles) each with a parametric value less than or equal to x. Thus $N(x)$ is necessarily discontinuous with increment 0 or a positive integer at each point x. This is perhaps the best reason why such processes are called point processes.

To fix our ideas, let us consider the simple Poisson process discussed in Section 3 of the previous chapter and let $N(t)$ denote the number of Poisson events occurring on the t-axis. In any infinitesimal interval $(t, t + \Delta)$ there is a jump of magnitude unity in $N(t)$ with probability $\lambda\Delta$ and no jump with probability $1 - \lambda\Delta$. Thus the contribution to Φ from Δ can be written as

$$(4.3) \quad (1 - \lambda\Delta) \, 1 + \lambda\Delta \, e^{i\theta(t)} + o(\Delta) = (1 - \lambda\Delta)(1 + \lambda\Delta \, e^{i\theta(t)}) + o(\Delta).$$

Since the Poisson events are independently distributed for disjoint intervals, the total contribution to Φ can be written as a product. Thus we have

$$(4.4) \qquad \Phi([\theta]) = \exp \{\lambda \int [\exp(i\theta(t)) - 1] \, dt\}.$$

In an attempt to estimate the stochastic fluctuations in age distribution, Kendall [2] defined the moment-generating functional by

$$(4.5) \qquad M[\theta(x); t] = E \{\exp \int_0^\infty \theta(x) \, dN(x, t)\}$$

where $N(x, t)$ is the number of individuals of age less than or equal to x at time t. He expanded the cumulant-generating functional defined by

$$(4.6) \qquad K([\theta(x)]; t) = \log M(\theta(x); t)$$

in the form

(4.7) $$K([\theta(x)]; t) = \int_0^\infty \theta(x)\, \alpha(x, t)\, dx + \tfrac{1}{2} \int_0^\infty [\theta(x)]^2\, \beta(x, t)\, dx$$

$$+ \tfrac{1}{2} \int_0^\infty \int_0^\infty \theta(x)\, \theta(y)\, \gamma(x, y)\, dx\, dy + \cdots .$$

where the cumulant functions α, β, and γ are defined by

(4.8) $$E\{dN(x, t)\} = \alpha(x, t)\, dx + o(dx)$$
$$\mathrm{var}\{dN(x, t)\} = \beta(x, t)\, dx + o(dx)$$
$$\mathrm{covar}\{dN(x, t)\, dN(y, t)\} = \gamma(x, y, t)\, dx\, dy + o(dx\, dy).$$

It is obvious that the moment-generating functional is nothing but the characteristic functional introduced by Bochner and Le Cam. In fact, Kendall has obtained a formal differential equation governing M using the evolutionary nature of the population process. The function $\alpha(x, t)$ can be identified with $\bar{n}(x)$ introduced by Bhabha [18] although in this case the variance of $dN(x, t)$ is not equal to the mean value of $dN(x, t)$. The utility of the characteristic functional in evolutionary stochastic point processes has been demonstrated by Bartlett and Kendall [35] who have dealt with specific problems in population growth and cosmic ray cascades.

REFERENCES

1. W. Feller, *An Introduction to Probability Theory and Its Applications*, John Wiley, New York, 1950.
2. D. G. Kendall, *J. Roy. Statist. Soc.*, B **11** (1949), 230.
3. M. S. Bartlett, *Stochastic Processes*, Cambridge, 1955.
4. J. E. Moyal, in *Symposium on Stochastic Processes, J. Roy. Statist. Soc.*, B **11** (1949).
5. H. Wold, in *Le Calcul des probabilités et ses applications*, C.N.R.S., Paris, 1949.
6. D. R. Cox, *J. Roy. Statist. Soc.*, B **17** (1955), 129.
7. W. L. Smith, *J. Roy. Statist. Soc.*, B **20** (1958), 243.
8. J. A. McFadden, *J. Roy. Statist. Soc.*, B **24** (1962), 364.
9. F. J. Beutler and O. A. Z. Leneman, *Act. Math.* **116** (1966), 159.
10. M. S. Bartlett, *Ann. Inst. H. Poincaré*, **14** (1954), 35.
11. M. S. Bartlett, *J. Roy. Statist. Soc.*, **25** (1963), 264.
12. J. E. Moyal, *Act. Math.*, **108** (1962), 1.
13. J. de Boer, *Rep. Prog. Phys.* (London), **12** (1949), 305.
14. A. Nordsieck, W. E. Lamb, and G. E. Uhlenbeck, *Physica*, **7** (1940), 344.

15. N. N. Bogoliubov, in *Studies in Statistical Mechanics*, Vol. I, North Holland Publishing Co., Amsterdam, 1962.
16. J. G. Kirkwood, *J. Chem. Phys.*, **14** (1946), 180.
17. M. Born and H. S. Green, *Proc. Roy. Soc.* (London), **A 189** (1947), 103.
18. H. J. Bhabha, *Proc. Roy. Soc.* (London), **A 202** (1950), 301.
19. A. Ramakrishnan, *Proc. Camb. Phil. Soc.*, **46** (1950), 595.
20. A. Ramakrishnan and T. K. Radha, *Proc. Camb. Phil. Soc.*, **57** (1961), 843.
21. S. K. Srinivasan and K. S. S. Iyer, *Nuovo Cimento*, **33** (1964), 273.
22. S. K. Srinivasan, in *Symposia in Theoretical Physics*, Vol. 4, Plenum Press, New York, 1967, p. 143.
23. D. G. Kendall, *Theory of Probability and Its Applications*, **9** (1964), 1.
24. L. Janossy, *Proc. Roy. Irish Acad. Sci.*, **A 53** (1950), 181.
25. S. K. Srinivasan, *Zast. Mat.*, **6** (1961–62), 209.
26. H. Messel, in *Progress in Cosmic Ray Physics*, Vol. II, North Holland Publishing Co., Amsterdam, 1954.
27. A. Ramakrishnan, *Proc. Camb. Phil. Soc.*, **48** (1951), 451.
28. W. T. Scott, *Phys. Rev.*, **82** (1951), 893.
29. W. T. Scott and G. E. Uhlenbeck, *Phys. Rev.*, **62** (1942), 497.
30. A. Ramakrishnan, *Stochastic Processes and Their Applications to Physical Problems*, Ph.D. Thesis, University of Manchester, 1951.
31. B. O. Koopman, *Proc. Nat. Acad. Sci. U.S.A.*, **36** (1950), 202.
32. S. Bochner, *Ann. Math.*, **48** (1947), 1014.
33. L. Le Cam, *C.R. Acad. Sci.* (Paris), **224** (1947), 710.
34. S. Bochner, *Harmonic Analysis and the Theory of Probability*, University of California Press, Berkeley, 1955.
35. M. S. Bartlett and D. G. Kendall, *Proc. Camb. Phil. Soc.*, **47** (1951), 65.

Chapter 3

POINT PROCESSES—PRODUCT DENSITIES

1. Introduction

Let us consider a stochastic process progressing with respect to a parameter t, the process itself being specified in relation to objects or entities each labeled with the random value of a continuous parameter x. The symbol x may stand for the energy of a cosmic ray particle or the age of an individual in a population or the velocity and position of a neutron in a fission process. In the previous chapter, we have dealt with "point distribution" by studying such objects described by a finite set of points $x_1, x_2, \ldots$ representing their "types." The present chapter will provide a complementary but extremely useful method of studying point processes through the introduction of the correlation functions known under the name product densities. The concept of product densities in its most general form was introduced by Ramakrishnan [1] during the course of his investigation of the fluctuation problem of electromagnetic cascades. The utility of the product densities in other contexts has also been amply demonstrated by Ramakrishnan [2] who has dealt with general stochastic processes associated with random points on a line.* The layout of the present chapter is as follows. We shall present the concept of product densities in Section 2 and explain how the moments and correlations of the number of entities with parametric values distributed over a given interval can be obtained, while in Section 3 we shall be concerned with the relationship between the product densities and the characteristic functional as well as the product-density-generating functional introduced by Kuznestov and Stratonovich. Section 4 will deal with some general properties of product densities governing Markovian as well as non-Markovian point processes. In the final section, we shall discuss a few applications in order to illustrate the power of the method of product densities particularly in non-Markovian situations. Although the specific examples discussed fall outside the scope of the subject matter of this monograph, the motivation to include this

* For a detailed account of the various situations where product density technique has been used, see Srinivasan [3].

material stemmed from the interest on the part of some probability theorists in using the special techniques of point processes in queuing theory (see Kendall [4]) and traffic flow (see discussion in reference [5]).

2. Product Densities

As in the earlier section, we shall assume that the process evolves with respect to t which may be continuous or discrete and the entities are distributed over the parametric space X.* The central quantity of interest in Ramakrishnan's theory is $dN(x, t)$ denoting the number of entities having parametric values in the interval $(x, x + dx)$ at the general parametric value corresponding to t. We next assume that the probability that there is an entity with a parametric value in $(x, x + dx)$ is proportional to dx, while the probability that there exists more than one with parametric values in $(x, x + dx)$ is of order smaller than dx. Thus if $p(n)$ is the probability that n entities have parametric values in $(x, x + dx)$, then

(2.1)
$$\begin{aligned} p(1) &= f_1(x, t)\, dx \ + \mathrm{o}\,(dx) \\ p(n) &= \mathrm{o}\,(dx) \qquad\qquad n > 1 \\ p(0) &= 1 - f_1(x, t) + \mathrm{o}\,(dx) \end{aligned}$$

and the moments of n are given by

(2.2) $\quad E\{n^m\} = E\{[dN(x, t)]^m\} = \Sigma n^m p(n)$

$$\begin{aligned} &= E\{dN(x, t)\} + \mathrm{o}\,(dx) \\ &= f_1(x, t)\, dx + \mathrm{o}\,(dx), \end{aligned}$$

thus proving that every one of the moments of the stochastic variable $dN(x, t)$ is equal to the probability that the stochastic variable takes the value unity. $N(x, t)$ has an interesting interpretation in that it can be taken to be the random variable representing the number of entities with parametric values $\leqslant x$. However, in view of the peculiar properties of $dN(x, t)$, $N(x, t)$ is a random function of points on the t-axis and of intervals or sets of intervals on the x-axis.

With the help of the above assumption, it is easy to obtain a formula for

* We shall take X to be a closed interval of the real line the results of this section being equally applicable when X is a rectangle of the p-dimensional Euclidean space. The results given in this section are also true when we replace the closed interval by a subset of the real line provided we interpret the integrals in the Lebesgue-Stieltjes sense.

the moments of the number of entities in a given interval $[a, b]$ of parametric space. Thus the mean number of entities is given by

$$(2.3) \qquad E\{[N(b, t) - N(a, t)]\} = E\left\{\int_a^b dN(x, t)\right\}$$

$$= \int_a^b E\{dN(x, t)\}$$

$$= \int_a^b f_1(x, t)\, dx$$

a relation which makes manifestly clear the idea that while $f_1(x, t)\, dx$ has a probability density interpretation, the integral of $f_1(x, t)$ over x yields only the mean number of entities in the range of integration. An equivalent method of interpreting such a situation due to Bartlett [6] consists in observing that only probability densities can be attached to particular values of x and not non-zero probabilities.

The mean square number of entities is formally given by

$$(2.4a) \qquad E\{[N(b, t) - N(a, t)]^2\} = \int_a^b \int_a^b E\{dN(x_1, t)\, dN(x_2, t)\}.$$

In view of the singular behavior of the random variable $dN(x, t)$ the line $x_1 = x_2$ in the domain of integration yields a non-negligible contribution to the integral. Accordingly we can split the domain of integration into two parts according as $x_1 \neq x_2$ and $x_1 = x_2$. Thus

$$(2.4b) \qquad E\{[N(b, t) - N(a, t)]^2\} = \int_a^b \int_a^b E\{dN(x_1, t)\, dN(x_2, t)\}_{x_1 \neq x_2}$$

$$+ \int_a^b \int_a^b E\{dN(x_1, t)\, dN(x_2, t)\}\delta(x_1 - x_2)$$

where δ is the Dirac delta function. Ramakrishnan has defined the product density of degree two by

$$(2.5) \qquad E\{dN(x_1, t)\, dN(x_2, t)\}_{x_1 \neq x_2} = f_2(x_1, x_2; t)\, dx_1\, dx_2$$

which can be interpreted as the simultaneous probability that an entity has a parametric value in $(x_1, x_1 + dx_1)$ and another in $(x_2, x_2 + dx_2)$ provided the two differential elements do not overlap, this being regardless of the entities assuming values in other ranges of x. Thus (2.5) can be simplified using (2.2)

$$(2.6) \quad E\{[N(b, t) - N(a, t)]^2\} = \int_a^b \int_a^b f_2(x_1, x_2; t)\, dx_1\, dx_2 + \int_a^b f_1(x)\, dx.$$

Equations (2.6) and (2.3) have also been obtained by Kendall in his study of the stochastic fluctuation in the age distribution. Kendall has defined the cumulant densities by

$$(2.7) \qquad \alpha(x, t)\, dx = \text{var } \{dN(x, t)\} = E\{dN(x, t)\}$$

$$(2.8) \qquad \gamma(x, y, t)\, dx\, dy = \text{cov } \{dN(x, t)\, dN(y, t)\}_{x \neq y}$$

from which we obtain

$$(2.9) \qquad \gamma(x, y, t) = f_2(x, y, t) - f_1(x, t) f_1(y, t).$$

The relations (2.1) have also been noticed by Kendall [7] who for this reason observes that n is asymptotically a Poisson variable, its mean and variance being equal (to first order in dx).

In a similar manner, we can define product densities of any higher order by

$$(2.10) \quad f_n(x_1, x_2, \ldots, x_n; t)\, dx_1\, dx_2 \cdots dx_n = E\{dN(x_1, t)\, dN(x_2, t) \\ \cdots dN(x_n, t)\}$$

representing the joint probability that there exists one entity having a parametric value in $(x_1, x_1 + dx_1)$, one having a parametric value in $(x_2, x_2 + dx_2)$ $\cdots$, and one having a parametric value in $(x_n, x_n + dx_n)$ provided the intervals $dx_1, dx_2, \cdots dx_n$ do not overlap. If, however, two of the dx's overlap, then the right-hand side of (2.10) becomes

$$(2.11) \quad E\{dN(x_1, t)\, dN(x_2, t) \cdots [dN(x_{n-1}, t)]^2\}$$

$$= E\{dN(x_1, t)\, dN(x_2, t) \cdots dN(x_{n-1}, t)\}$$
$$= f_{n-1}(x_1, x_2, \ldots, x_{n-1}; t)\, dx_1\, dx_2 \cdots dx_{n-1}$$

where we have assumed $dx_{n-1} = dx_n$. The degeneracy to a lower order product density is due to the fact that every one of the moments of $dN(x, t)$

is equal to the probability that the random variable $dN(x, t)$ takes the value 1. There may be more general situations wherein such a simple result as (2.11) may not be possible and this will be discussed in Chapter 4 when we deal with the extensions of product-density techniques.

To obtain the r-th moment of the number of entities with parameters distributed over the interval $[a, b]$, we proceed in a manner analogous to (2.3) and (2.4) and take into account the degeneracies that will arise due to the overlap of the differential elements of the parametric space. Thus we can write

$$(2.12) \qquad E\{[N(b, t) - N(a, t)]^r\} = \sum C_s^r \int_a^b \int_a^b \cdots \int_a^b$$
$$f_s(x_1, x_2, \ldots, x_s; t)\, dx_1\, dx_2 \cdots dx_s$$

where the coefficient C_s^r denoting the number of $(r - s)$-fold degeneracy in an r-fold product is a function of r and s only and in no way dependent on the product density f_s. To calculate C_s^r we consider the special case when the total number of entities are fixed, say N. The product density of degree one can be written as

$$(2.13) \qquad f_1(x, t)\, dx = N f_1^0(x, t)\, dx$$

with the condition

$$(2.14) \qquad \int f_1^0(x, t)\, dx = 1$$

where $f_1^0(x, t)\, dx$ represents the probability that a particular entity has a parametric value between x and $x + dx$. In a similar manner, we find

$$(2.15)\ f_s(x_1, x_2, \ldots, x_s; t)$$
$$= N(N - 1) \cdots (N - s + 1) f_1^0(x_1, t) f_1^0(x_2, t) \cdots f_1^0(x_s\, t)$$
$$(s = 2, 3 \ldots, N).$$

Substituting (2.15) in (2.12) and integrating over the entire x-space, we get

$$(2.16) \qquad N^r = \sum C_s^r N(N - 1) \cdots (N - s + 1).$$

The coefficients are obtained by taking $N = 1, 2, \ldots$ successively.

The product-density technique can also be used to obtain the correlations of the number of entities corresponding to different parametric intervals. For convenience, let us define $N(y, t)$ by

$$(2.17) \qquad N(y, t) = \int_0^y dN(x, t).$$

The second order correlation is given by

$$(2.18) \quad E\{N(y_1, t) \, N(y_2, t)\} = \int_0^{\min(y_1, y_2)} f_1(x, t) \, dx + \int_0^{y_1} \int_0^{y_2} f_2(x_1, x_2; t) \, dx_1 \, dx_2$$

while the correlation of degree three is given by

$$(2.19) \quad E\{N(y_1, t) \, N(y_2, t) \, N(y_3, t)\}$$

$$= \int_0^{\min(y_1, y_2, y_3)} f_1(x, t) \, dx + \left[\int_0^{\min(y_1, y_2)} dx_1 \int_0^{y_3} + \int_0^{\min(y_1, y_3)} dx_1 \int_0^{y_2} \right.$$

$$\left. + \int_0^{\min(y_2, y_3)} dx_1 \int_0^{y_1} \right] f_2(x_1, x_2; t) \, dx_2$$

$$+ \int_0^{y_1} dx_1 \int_0^{y_2} dx_2 \int_0^{y_3} dx_3 \, f_3(x_1, x_2, x_3; t).$$

It is to be noted that the coefficients C_s^r do not appear at all. On the other hand, the different terms that go into a single partition corresponding to a given order of degeneracy break up into an equal number of integrals, the domain of integration being different in each case. Similar equations can be written down for higher order correlations and it can be easily verified that on making $t_1 = t_2 = \cdots = t_n$, these equations will reduce to the moment formula given by (2.12).

Before we proceed to discuss the properties of the product densities, we wish to make a few remarks about the contributions from the Russian school. Kuznestov and Stratonovich [8], apparently unaware of the main results of the point processes, have reintroduced the product densities by an ingenious use of combinatorial arguments similar to the ones used by Bhabha and Ramakrishnan during the primitive stage of the development of the theory of product densities (see reference [9]). They have succeeded in expressing the Janossy density in terms of the product densities, a result which is of great consequence in physical application. Kuznestov, Stratonovich, and Tikhonov [10] have made systematic use of these techniques of point processes in noise and optimum filtration.

3. Characteristic and Product-Density-Generating Functional

The product densities are very intimately connected to the characteristic functional introduced in Chapter 2. The appropriate characteristic functional in the present case is given by

$$(3.1) \qquad \mathscr{L}([q], t) = E \{\exp i\!\int\! q(x) \, dN(x, t)\}$$

Expanding the exponential and inverting the order of summation and expectation symbol, we find

$$(3.2) \qquad \mathscr{L}([q], t) = \Sigma \frac{i^m}{m!} \int\!\int \cdots \int q(x_1) \, q(x_2) \cdots q(x_m)$$

$$E \{dN(x_1, t) \, dN(x_2, t) \cdots dN(x_m, t)\}.$$

Next, we consider the number of ways in which confluences of x_i's occur in the integrand. In the simple problem of moments the number of ways in which $(m - h)$-fold confluence can occur has been found to be C_h^m. However, in the above expression, we have to break up into a number of "complexions" in view of the presence of the product $q(x_1) \, q(x_2) \cdots q(x_m)$ in the integrand, a complexion being characterized by the set of numbers $1_1, 1_2, \ldots, 1_h$ such that $1_1 + 1_2 + \cdots + 1_h = m$. Denoting the number of ways of obtaining a typical complexion ν belonging to (h, m) by $C_h^m(\nu)$ we find

$$(3.3) \qquad \sum_\nu C_h^m(\nu) = C_h^m$$

the summation over ν indicating the sum over all possible distinct complexions. Thus (3.2) can be written in the form

$$(3.4) \qquad \mathscr{L}([q], t) = 1 + \sum_{m=1}^{\infty} \frac{i^m}{m!} \sum_h \sum_\nu C_h^m(\nu)$$

$$\int dx_1 \int dx_2 \cdots \int dx_h \, f_h(x_1, x_2, \ldots, x_h; t)$$

$$[q(x_1)]^{l_1} \, [q(x_2)]^{l_2} \cdots [q(x_h)]^{l_h}$$

where the coefficient $C_h^m(\nu)$ can be obtained by combinatorial arguments by observing that it is the number of ways in which m distinguishable objects can be thrown into h groups of $l_1, l_2 \ldots, l_h$. Thus if only p out of the h

integers $l_1, l_2, \ldots, l_h$ are different and $k_1, k_2, \ldots, k_p$ are the orders of the degeneracies in each of them, we have

$$(3.5) \qquad C_h^m(\nu) = \frac{m!}{l_1! \, l_2! \cdots l_h! \, k_1! \, k_2! \cdots k_p!}.$$

From (3.4) it is easy to identify the product densities by

$$(3.6) \qquad (-i)^h \left. \frac{\delta^h \mathcal{L}([q], t)}{\delta q(x_1) \, \delta q(x_2) \cdots \delta q(x_h)} \right|_{q=0} = f_h(x_1, x_2, \ldots, x_h; t)$$

provided $x_1 \neq x_2 \neq \cdots \neq x_h$. In the more general case when some of x_i's are equal, there will be additional singular terms containing products of lower order product densities and appropriate delta functions involving x_i's.

We next wish to establish the connection between product densities and the Janossy densities introduced in Chapter 2. Suppose we integrate a product density of degree h with respect to only one variable. Then

$$(3.7) \quad \int_{x_h} f_h(x_1, x_2, \ldots, x_h; t) \, dx_1 \, dx_2 \cdots dx_h$$
$$\equiv S_{h-1}(x_1, x_2, \ldots, x_{h-1}; t) \, dx_1 \, dx_2 \cdots dx_{h-1}$$

represents the expectation value of the product of the number of entities in the range of integration of x_h and the number of entities in $dx_1 \cdots dx_{h-1}$. Thus $S_{h-1}(x_1, x_2, \ldots, x_{h-1}; t)$ is no longer a product density nor does $S_{h-1}(x_1, x_2, \ldots, x_{h-1}; t) \, dx_1 \, dx_2 \cdots dx_{h-1}$ represents a probability magnitude. However, we consider a special case in which the total number of entities are fixed. Denoting by $f_h^{(N)}(x_1, x_2, \ldots, x_N; t)$ the corresponding product density, we find

$$(3.8) \qquad S_{h-1}^{(N)}(x_1, x_2, \ldots, x_{h-1}; t) = \int_{x_h} f_h^{(N)}(x_1, x_2, \ldots, x_h; t) \, dx_h$$
$$= (N - h + 1) f_{h-1}^{(N)}(x_1, x_2, \ldots, x_{h-1}; t)$$

where the integration over x_h is performed over the entire range. Such an interesting result is due to the fact that the number of entities outside the ranges $dx_1, dx_2 \ldots, dx_{h-1}$ is $(N - h + 1)$ conditional upon the existence of one in dx_1, one in dx_2, and one in dx_{h-1}. Using (3.8), we can obtain

$$(3.9) \qquad f_h^{(N)} = \frac{1}{(N-h)!} \int_{x_{h+1}} \int_{x_{h+2}} \cdots \int_{x_N} f_N^{(N)}(x_1, x_2, \ldots, x_N, t)$$
$$dx_{h+1} \, dx_{h+2} \cdots dx_N$$

an interesting result which leads to the relation

$$(3.10) \qquad J_h(x_1, x_2, \ldots, x_h; t) = P(h, t) f_h^{(h)}(x_1, x_2, \ldots, x_h; t)$$

where $P(h, t)$ represents the probability that there are h entities in the entire range of x. Observing that

$$(3.11) \qquad \int\limits_{x_1} \int\limits_{x_2} \cdots \int\limits_{x_h} f_h^{(h)}(x_1, x_2, \ldots, x_h; t)\, dx_1\, dx_2 \cdots dx_h = h!$$

we find

$$(3.12) \quad P(h) = \left(\frac{1}{h!}\right) \int\limits_{x_1} \int\limits_{x_2} \cdots \int\limits_{x_h} J_h(x_1, x_2 \ldots, x_h; t)\, dx_1\, dx_2 \cdots dx_h$$

a result that has been obtained in Chapter 2 by the direct use of the properties of J_h.

From the definition of product densities, it follows that

$$(3.13) \qquad f_h(x_1, x_2, \ldots, x_h; t) = \sum_{n=h}^{\infty} P(n) f_h^{(n)}(x_1, x_2, \ldots, x_h; t).$$

If we use (3.13), we obtain

$$(3.14)\ f_h(x_1, x_2, \ldots, x_h; t) = \sum_{n=h}^{\infty} \frac{1}{(n-h)!} \int\limits_{x_{h+1}} \int\limits_{x_{h+2}} \cdots \int\limits_{x_N}$$
$$J_N(x_1,\ x_2 \ldots, x_N; t)\, dx_{h+1}\, dx_{h+2} \cdots dx_N$$

a result that was promised in earlier sections. As we have demonstrated in Section 3 of Chapter 2, equation (3.14) was the starting point for the work of Bhabha [11] who expressed the Janossy densities in terms of saw-tooth functions. In fact, the same technique has been used by Srinivasan [12] for the extension of product densities to include multiple points and this will be discussed in the next chapter. The above relationship also explains why the product densities enjoy a privileged position in the calculation of moments. The inverse formula corresponding to (3.14) has been noted by Kuznestov and Stratonovich [8] (see also reference [13])

$$(3.15) \qquad J_n(x_1, x_2, \ldots x_n; t) = \sum_{k=n}^{\infty} \frac{(-)^{k-n}}{(k-n)!} \int\limits_{x_{n+1}} \int\limits_{x_{n+2}} \cdots \int\limits_{x_k}$$
$$f_k(x_1, x_2, \ldots, x_k; t)\, dx_{n+1}\, dx_{n+2} \cdots dx_k,$$

If we integrate both sides of (3.15) over $x_1, x_2, \ldots, x_n$ covering the entire range, we obtain

$$(3.16) \qquad P(n, t) = \frac{1}{n!} \sum_{k=n}^{\infty} \frac{(-)^{k-n}}{(k-n)!} \int_{x_1} \int_{x_2} \cdots \int_{x_k}$$

$$f_k(x_1, x_2, \ldots, x_k; t) \, dx_1 \, dx_2 \cdots dx_k.$$

Thus if $G(u, t)$ is the probability generating function of $P(n, t)$ then it follows

$$(3.17) \quad G(u, t) = 1 + \sum_{k=1}^{\infty} \frac{(u-1)^k}{k!} \int_{x_1} \int_{x_2} \cdots \int_{x_k} f_k(x_1, x_2, \ldots, x_k; t)$$

$$dx_1 \, dx_2 \cdots dx_k$$

a result that is of great importance in practical applications.

Kuznestov and Stratonovich have also introduced the product-density-generating functional by

$$(3.18) \qquad L([q], t) = \sum_{n=0}^{\infty} \frac{1}{n!} \int_{x_1} \int_{x_2} \cdots \int_{x_n} f_n(x_1, x_2, \ldots, x_n; t)$$

$$q(x_1), q(x_2) \cdots q(x_n) \, dx_1 \, dx_2 \cdots dx_n$$

so that the product densities can be obtained by taking functional derivatives of L. It is interesting to compare (3.18) with (3.6). While the functional derivatives of $\mathscr{L}$ and L are equal whenever $x_1 \neq x_2 \neq \cdots \neq x_h$, the derivatives of the characteristic functional exhibit singularities of the delta function type along $x_1 = x_2 \cdots = x_h$. In view of this, it may be convenient to deal with L rather than $\mathscr{L}$ in practical situations. Kuznestov and Stratonovich have also introduced a different type of correlation functions which may be rightly called actual correlation functions. These are nothing but generalizations of Kendall's cumulant functions to higher order. There are a number of ways by which these functions can be introduced, the simplest one being through the product density-generating-functional:

$$(3.19) \qquad L([q], t) = \exp \sum_{n=1}^{\infty} \frac{1}{n!} \int_{x_1} \int_{x_2} \cdots \int_{x_n} g_n(x_1, x_2, \ldots, x_n; t)$$

$$u(x_1) \, u(x_2) \cdots u(x_n) \, dx_1 \, dx_2 \cdots dx_n.$$

Thus $g_n(x_1, x_2, \ldots, x_n; t)$ can be related to the product densities by taking the functional derivative of both sides of (3.19):

$$(3.20) \qquad f_1(x, t) = \left. \frac{\delta L}{\delta q(x)} \right|_{q=0}$$

$$= g_1(x; t)$$

$$(3.21) \qquad f_2(x_1, x_2; t) = \left. \frac{\delta^2 L}{\delta q(x_1)\, \delta q(x_2)} \right|_{q=0}$$

$$= g_1(x_1; t)\, g_1(x_2; t) + g_2(x_1, x_2; t)$$

$$(3.22) \qquad f_3(x_1, x_2, x_3; t) = \left. \frac{\delta^3 L}{\delta q(x_1)\, \delta q(x_2)\, \delta q(x_3)} \right|_{q=0}$$

$$= g_1(x_1; t)\, g_1(x_2; t)\, g_1(x_3; t)$$
$$+ g_1(x_1; t)\, g_2(x_2, x_3; t)$$
$$+ g_1(x_2; t)\, g_2(x_3, x_1; t)$$
$$+ g_1(x_3; t)\, g_2(x_1, x_2; t)$$
$$+ g_3(x_1, x_2, x_3; t).$$

For a Poisson process, it will be shown presently in Section 4 that $g_n(x_1, x_2, \ldots, x_n; t)$ $(n > 1)$ vanishes so that this justifies the name "actual correlation function."

4. Some General Properties of Product Densities

In this section, we shall attempt to describe some general properties of the point processes by considering Poisson processes and renewal processes. Although the theory of point processes emanated from simple stochastic processes associated with random points on a line, of late there have been innumerable physical situations where the underlying space is no longer Euclidean. Although such examples of point processes have been observed in a wide spectrum of phenomena there has been not much of a systematic study of the theory of point processes, and the mathematical treatments (see, for example, Beutler and Leneman [14]) are confined only to those processes originally considered by Wold [15]. Bartlett [6] has adopted a phenomenological definition which includes many of the features mentioned above. The recent work of Moyal [16] deserves special mention and he has put many of the phenomenological results on a sound mathematical footing. However, there still remains the main task of enumerating the properties of point processes and this is best done by discussing the properties of product densities. We hope that what is discussed below, though fragmentary in nature, will bring to focus the intricate nature of point processes in general and also the need for further efforts in this direction.

4.1. POISSON PROCESS

Let us consider the special case when the product density of degree n is given by

$$(4.1) \qquad f_n(x_1, x_2, \ldots, x_n) = f_1(x_1) f_1(x_2) \cdots f_1(x_n)$$

corresponding to the situation when there is no correlation between the entities distributed in different parametric intervals.* If we now substitute (4.1) on the right-hand side of (3.17), and take the range of integration to be $(0, X)$, we obtain

$$(4.2) \qquad G(u) = \exp\left[(u - 1) \int_0^x f_1(x_1)\, dx_1\right]$$

a result which proves that the underlying process is an inhomogeneous Poisson process with parameter $f_1(x)$. Thus a Poisson distribution of entities can be characterized by (4.1) and hence by the condition

$$(4.3) \qquad g_n(x_1, x_2, \ldots, x_n) = 0 \qquad n > 1$$

where g_n is the actual correlation function introduced in the previous section.

4.2. YULE-FURRY PROCESS

Another interesting result can be obtained by modifying (4.1) as

$$(4.4) \qquad f_n(x_1, x_2, \ldots, x_n) = n!\, f_1(x_1) f_1(x_2) \cdots f_1(x_n)$$

which leads to

$$(4.5) \qquad G(u) = \left[1 - (u - 1) \int_0^x f_1(x_1)\, dx_1\right]^{-1}$$

$$(4.6) \qquad P(n) = \frac{[N(X)]^n}{[1 + N(X)]^{n+1}}$$

* We have suppressed the parameter t as the dependence on such a parameter is of no significance in the present case.

where $N(X)$ is the mean number of entities with parametric values in $[0, X]$. Thus (4.4) characterizes the inhomogeneous Yule-Furry process generated by a single entity. It is instructive to derive (4.4) by assuming that the process evolves with respect to x (x being the time parameter) and that at $x = 0$ there is a single primary entity. We further assume that the primary entity generates a population of entities and that the probability that any particular entity produces another during the time interval $(x, x + dx)$ is $\lambda\, dx$ the probability of production of more than one being of a smaller order of magnitude than dx. Then $f_1(x)$ satisfies the equation

$$(4.7) \qquad f_1(x) = \lambda + \lambda \int_0^x f_1(y)\, dy.$$

The above equation is obtained by arguing that the entity that is produced between x and $x + dx$ is produced either by the primary (this happens with probability $\lambda\, dx$) or by a secondary which, in turn, must have been produced sometime between 0 and x. In a similar manner, we obtain

$$(4.8) \qquad f_2(x_1, x_2) = 2f_1(x_1) + \lambda \int_0^{x_2} f_2(x_1, y)\, dy$$

$$(4.9) \qquad f_n(x_1, x_2, \ldots, x_n) = nf_{n-1}(x_1, x_2, \ldots, x_{n-1})$$

$$+ \int_0^{x_n} \lambda f_n(x_1, x_2, \ldots, x_{n-1}, y)\, dy.$$

These equations can be solved by elementary methods (see Ramakrishnan and Srinivasan [17]) and we recover (4.4) with the condition

$$(4.10) \qquad f_1(x) = e^{\lambda x}\lambda$$

so that the mean number $N(x)$ is given by

$$(4.11) \qquad N(x) = e^{\lambda x} - 1.$$

Equation (4.11) yields only the mean number of entities produced between 0 and x, the total number at time x being $e^{\lambda x}$. In this case, the higher order actual correlation functions do not vanish as is to be expected.

In this process, once the entities are created they remain. If we introduce constant probability of the annihilation, we obtain the homogeneous birth and death process. In such a situation, equations (4.7) through (4.9) require drastic modification and this will be discussed in detail in Chapter 9.

4.3. RENEWAL PROCESSES

Let us consider a special case in which we identify the entities with certain types of events occurring along the time axis. Then the process is uniquely defined by the family of product densities $f_n(x_1, x_2, \ldots, x_n)$ $(n = 1, 2, \ldots)$ where $f_n(x_1, x_2, \ldots, x_n) \, dx_1, dx_2, \ldots, dx_n$ represents the probability that one event occurs in $(x_1, x_1 + dx_1)$, one in $(x_2, x_2 + dx_2) \ldots$, and one in $(x_n, x_n + dx_n)$ irrespective of the events that occur elsewhere. If we further assume that the process is stationary, then it follows

(4.12) $$f_1(x) = \text{a constant} = \alpha$$

(4.13) $$f_2(x_1, x_2) = \alpha h_1(x_2 - x_1)$$

(4.14) $$f_n(x_1, x_2, \ldots, x_n) = \alpha h_{n-1}(x_2 - x_1, x_3 - x_1, \ldots, x_n - x_1).$$

A knowledge of these correlation functions will enable us to predict all the statistical properties of the process.

Renewal processes can be considered as a special case of the stationary point processes defined by equations (4.12) through (4.14). To achieve this, let us consider a more general case in which the process is switched on at $x = 0$ and that an event has occurred at $x = 0$. The renewal process can be characterized in two equivalent modes. The first due to Ramakrishnan [9] (see also reference [18]) consists in defining the higher order product densities by

(4.15) $$f_n(x_1, x_2, \ldots, x_n) = f_1(x_1) f_1(x_2 - x_1) \cdots f_1(x_n - x_{n-1})$$
$$n = 2, 3, \ldots.$$

The above relation is equivalent to stating that given an event has occurred at $x = 0$, the probability that an event occurs between x and $x + dx$ (whether or not an event occurred between 0 and x) is determined by the function $f_1(x)$ and is independent of what happened before $x = 0$. From this, it also follows that given an event has occurred at $x = 0$, the probability that the *next event* occurs between x and $x + dx$ is also independent of what happened before $x = 0$. In fact, this is exactly the manner in which a renewal process has been originally defined (see, for example, Cox and Smith [19]). The product density method of description is decidedly superior to the conventional mode of description since all the relevant information regarding the renewal process can be readily obtained from the product densities. We shall demonstrate the advantages that generally accrue from the theory of product densities when we deal with some simple applications of the product density technique in the next section.

In the previous paragraph, we have assumed that an event occurs at $x = 0$. This condition can be relaxed by assuming that the origin is chosen arbitrarily in which case (4.15) is replaced by

$$(4.16) \quad f_n(x_1, x_2, \ldots, x_n) = f(x_1) f_1(x_2 - x_1) f_1(x_3 - x_2) \cdots f_1(x_n - x_{n-1})$$

where $f(x)$ is, in general, different from $f_1(x)$. It can be shown that either of the two functions $f(x)$ and $f_1(x)$ determines and is determined by the distribution of the length of the interval between two successive events. In fact, this has been done by Cox [20] who used slightly different terminology and notation. The renewal density and the modified renewal density can be identified to be $f_1(x)$ and $f(x)$ respectively.

We wish to conclude this short discussion on renewal processes by obtaining an explicit expression for the probability frequency function of the total number of renewals under a special circumstance. Let us assume that we start with an event at $x = 0$ and that

$$(4.17) \qquad\qquad \int_0^\infty f_1(x)\, dx = \alpha < \infty$$

so that the mean of the number of renewals over an infinite period is finite. If we substitute (4.15) into (3.17) and observe that there are $k!$ terms arising out of ordering the x_i's in the integrand, we obtain

$$(4.18) \qquad\qquad G(u) = \frac{1}{1 - (u - 1)\alpha}$$

which can be identified to be the generating function corresponding to the Yule-Furry process. Equation (4.18) does not seem to have been noticed by renewal theorists.

4.4. RECURRENT EVENTS—DISCRETE PRODUCT DENSITIES

Bartlett (see [6], pages 56–59) has introduced certain functions to facilitate the study of recurrence and passage times for the study of renewal processes. The object of this subsection is to identify them as product densities. Suppose we wish to study the recurrence of a particular state S in a random sequence, given that S occurred at time zero. We shall assume time to be discrete and introduce the random variable e_r corresponding to the time label r taking the values 1 and 0 according to whether the state of the system is S or not. Thus the probability that the system is found in state S at any subsequent time r given that it was found in state S at time zero is given by

$$(4.19) \qquad P_r(S_r|S_0) = \{e_r\}$$

where the curly brackets denote the expectation value of the quantity within the bracket. Likewise we can introduce the random variable $\overline{e_r}$ which takes the value 0 and 1 according to whether the system is in the state S or not. Then the probability that the system is in a state other than S at time r is given by

$$(4.20) \qquad P_r(\text{not } S_r|S_0) = \overline{\{e_r\}} = 1 - \{e_r\}.$$

The probability of the system being found in state S at times r and s is given by

$$(4.21) \qquad P_r(S_r, S_s|S_0) = \{e_r\, e_s\}.$$

Bartlett has shown that the probability of the first recurrence at time r is given by

$$(4.22) \quad \{\overline{e_1}\, \overline{e_2} \cdots \overline{e_{n-1}}\, e_n\} = \{e_n\} - \sum_{r=1}^{n-1} \{e_r\, e_n\} +$$
$$\sum_{r=1}^{n-2} \sum_{s=r+1}^{n-1} \{e_r\, e_s\, e_n\} - \cdots$$

where the curly brackets of three or more e's have an interpretation similar to (4.21). Based on (4.22) and (4.19), Bartlett has discussed the recurrence and first passage times of renewal processes. He has also obtained an analogous result for the continuous time.

It is interesting to observe that (4.19) defines the product density of degree one over the discrete parametric space. If we sum e_r over r, we obtain the mean number of recurrences of the state S. Likewise equation (4.21) can be identified to be the product density of degree two. The left-hand side of equation (4.22) can be interpreted as the probability of exactly one recurrence of the state at time n and none elsewhere (provided we confine our attention to times $\leq n$). Thus (4.22) provides the explicit relation between the Janossy density of degree one and the product densities of various orders. In fact, the continuous analogue of this result due to Bartlett is nothing but the Kuznestov-Stratonovich formula (3.15) of the previous section for the special case $n = 1$. From a conceptual and computational point of view the identification of the functions $\{e_r\}, \{e_r\, e_s\}, \ldots$ as product densities will introduce enormous simplification in the understanding of recurrence and passage times.

Finally, we wish to observe that equation (4.22) and its continuous version make manifestly clear the inherent difficulty of the general problem of recurrence times.

4.5. NON-MARKOVIAN PROCESSES

Point processes that are non-Markovian in their character are difficult to deal with. Of course, such processes are defined by the chain of product densities of all orders. If we impose stationarity, a slight simplification would result and this has been expressed by equations (4.12) through (4.14). Equation (4.13) is useful in the determination of the power spectrum which is nothing but the Fourier transform of $f_2(x_1, x_2)$. Such a result is extremely valuable if power spectrum is available through an experiment, in which case the function $h_1(x)$ can be determined. Apart from this, it is not clear as to how one could proceed in a general situation. No doubt, some general relations connecting the moments of the number of entities (these are very often taken to be points on a line) and the product densities may be obtained (see, for example, McFadden [21], McFadden and Weissblum [22], Beutler and Lenemann [14]). The implications of these results in any physical situation are not clear partly because the product densities have not been used in the proper context. However, the recent work of McFadden [23] is a significant departure from the conventional treatment of point processes. He has demonstrated the utility of the product densities in a non-Poissonian context and much remains to be done in this direction.

In some special cases, it is indeed possible to obtain higher order product densities even though the basic process is non-Markovian. This is usually done by imbedding the non-Markovian process in a suitable Markov process. Such a method of viewing the given non-Markovian process as the projection of a Markov process appears to be fairly well known. Wang and Uhlenbeck [24] have used such methods to study the Brownian motion of coupled oscillators. The author [25] has used an exactly similar procedure for the study of integrals of a class of random functions (see also Mathews and Srinivasan [26]). Cox [27] appears to have used similar methods by introducing supplementary variables, and these results are expected to be significant in queuing theory (see, for example, Kendall [4]). In the next section, we shall illustrate the method of obtaining the higher order product densities in shot noise theory where the arrival of an electron in the anode at a point on the time axis influences subsequent arrivals.

5. Applications of Product Densities

Since a considerable part of this monograph is to deal mainly with the applications of the product densities and their extensions, it might seem puzzling to the reader as to why a section under this heading is included. As we have mentioned in the Introduction as well as in the last few paragraphs of the previous section, the development of this subject is nascent and many of the properties of these densities which, in our opinion are fairly important, cannot be brought out in a vivid manner except through recourse to specific examples. Particularly the non-Markovian process, by its very definition being residuary in character, is fairly difficult to deal with in its full generality. However, if we restrict ourselves to specific physical processes or pheno-menological models, it is indeed possible to derive some of their properties. In the next few paragraphs, we shall present the physical problem and discuss the present state to the best of our knowledge.

5.1. NON-MARKOVIAN MODEL OF SHOT NOISE

Let us consider a stochastic process of the inhomogeneous Poisson type with the parameter $\lambda(t)$ where $\lambda(t)$ is a non-negative, continuous, and bounded function of t, t being the parameter with respect to which the process progresses. The probability that an event happens between t and $t + dt$ is $\lambda(t)\,dt$ while the probability that n events occur between 0 and t is given by

(5.1) $$P(n, t) = e^{-\Lambda(t)}\,[\Lambda(t)]^n/n!.$$

This type of Poisson process can be used to describe a number of physical phenomena such as electron emission in a counter and age dependent birth and death process (see, for example, Takacs [28]). In all these processes the parameter $\lambda(t)$ depends only on t and does not depend on either the number of events that have occurred before t or the points at which the events have occurred. Processes for which $\lambda(t)$ depends on the number of events that have occurred between 0 and t have received some attention (see Bartlett [6]). However, there are quite a number of physical phenomena wherein $\lambda(t)$ depends not only on the number of events that have occurred prior to t but on the positions of t also. An example of this kind of process is provided by the shot effect (see Rowland [29]) where an electron arriving at the anode between t_1 and $t_1 + dt_1$ diminishes the probability of any further arrival at a later time t_2 by $be^{-a(t_2 - t_1)}$ where a and b are some positive physical constants depending on the system. The process is switched on at $t = 0$ when the probability of the occurrence of an event is λ_0 (a constant). Thus the first event happens between t_1 and $t_1 + dt_1$ with probability $e^{-\lambda_0 t_1}\lambda_0 dt_1$ while the probability of occurrence of the second event between t_2 and $t_2 + dt_2$ is given by*

$$(5.2) \quad P_2(t_1, t_2)\, dt_2 = \left[\exp\left(-\int_{t_1}^{t_2} \{\lambda_0 - b \exp\left[-a(t' - t_1)\right]\}\, dt' \right) \right]$$
$$\{\lambda_0 - b \exp[-a(t_2 - t_1)]\}\, dt_2.$$

Let us denote the parameter characterizing the process by $\lambda(t)$ in analogy with the inhomogeneous Poisson process. The parameter $\lambda(t)$ *is no longer a deterministic* function of t but depends on the various random values of t at which the events have occurred. A typical realized value of $\lambda(t)$ corresponding to the events that have occurred at $t_1, t_2 \cdots t_n$ is given by

$$(5.3) \qquad \lambda^R(t) = \lambda_0 - b \sum_{i=1}^{n} \exp - a(t - t_i).$$

The probability measure corresponding to the above realized value can be calculated using (5.3). In the case of shot noise, the individual pulses signifying the arrivals of electrons in the anode give rise to a response $\phi(t)$ at time t after the occurrence of the pulse at $t = 0$ and we are interested in the cumulative response $r(t)$ given by

$$(5.4) \qquad r(t) = \sum_{i} \phi(t - t_i)$$

where t_i is the time of occurrence of the i-th pulse and vanishes for the negative values of its argument. It is easy to see that in the notation of Section 2, the response can be expressed as a stochastic integral

$$(5.5) \qquad r(t) = \int_{0}^{t} \phi(t - \tau)\, dN(\tau).$$

The mean value and the second order correlation of $r(t)$ are given by

$$(5.6) \qquad \overline{r(t)} = \int_{0}^{t} \phi(t - \tau) f_1(\tau)\, d\tau$$

* $P_2(t_1, t_2)$ is a conditional probability frequency function.

$$(5.7) \qquad \overline{r(t_1)\, r(t_2)} = \int_0^{t_1} d\tau_1 \int_0^{t_2} d\tau_2\, \phi(t_1 - \tau_1)\, \phi(t_2 - \tau_2) f_2(\tau_1, \tau_2)$$

$$+ \int_0^{\min(t_1, t_2)} \phi(t_1 - \tau)\phi(t_2 - \tau) f_1(\tau)\, d\tau$$

where f_1 and f_2 are the product densities of degree one and two of events occurring along the t-axis. Equation (5.7) is derived using arguments very similar to the derivation of (2.18) and (2.19). Thus the problem reduces to that of the determination of the product densities of the first two orders. The problem was first tackled by Rowland who obtained explicit expressions for the two moments of $r(t)$. However, there are some limitations in the final results obtained by Rowland. During the process of integration Rowland ignored the probability of $\lambda(t)$ dropping down to negative values. In fact, this is an important step in the computation of the probability frequency function of $\lambda(t)$. The author has given [30, 31] a complete solution to the problem by obtaining explicitly the product densities of the first three orders. We shall briefly sketch the important steps that lead to the determination of the product densities. For full details, the reader is referred to the original paper.

If $\pi(\lambda, t)$ is the probability frequency function of $\lambda(t)$, $\pi(\lambda, t)$ satisfies the equation

$$(5.8) \qquad \frac{\partial \pi(\lambda, t)}{\partial t} = (a - \lambda)\pi(\lambda, t) - a(\lambda_0 - \lambda)\frac{\partial \pi(\lambda, t)}{\partial \lambda}$$

$$+ (\lambda + b)\pi(\lambda + b, t).$$

Equation (5.8) is true only if $\lambda > 0$. When $\lambda < 0$, it is easy to see that $\pi(\lambda, t)$ satisfies the equation

$$(5.9) \qquad \frac{\partial \pi(\lambda, t)}{\partial t} = a\pi(\lambda, t) - a(\lambda_0 - \lambda)\frac{\partial \pi(\lambda, t)}{\partial \lambda}$$

$$+ (\lambda + b)\pi(\lambda + b, t).$$

However, in such a case $\lambda\, dt$ cannot be a probability magnitude. The difficulty can be overcome by defining λ' by

$$(5.10) \qquad \lambda' = \lambda \text{ for } \lambda > 0$$
$$= 0 \text{ otherwise.}$$

We observe that it is λ' that has probability significance and in any problem we have to deal with only the moments of λ'. Equation (5.8) is very difficult to solve for $\pi(\lambda, t)$ explicitly. However, it is possible to obtain the moments of λ'. Defining

$$(5.9) \qquad p(n, t) = \int_{-b}^{\infty} \pi(\lambda, t)\lambda'^n \, d\lambda$$

we obtain

$$\frac{\partial p(n, t)}{\partial t} = -nap(n, t) + na\lambda_0 p(n - 1, t)$$
$$+ \sum_{i=1}^{n} \binom{n}{i} p(n - i + 1, t)(-b)^i$$

with the conditions

$$(5.11) \qquad p(0, t) = 1, \qquad p(n, 0) = \lambda_0^n.$$

The first few moments can be explicitly calculated:

$$(5.12) \qquad p(1, t) = \frac{a\lambda_0}{(a + b)} + \frac{b\lambda_0}{a + b} e^{-(a+b)t}$$

$$(5.13) \qquad p(2, t) = \frac{(2\lambda_0 - a - 2b)b^2\lambda_0}{2(a + b)^2} e^{-2(a+b)t}$$
$$+ \frac{b\lambda_0(2a\lambda_0 + b^2)}{(a + b)^2} e^{-(a+b)t} + \frac{a\lambda_0(2a\lambda_0 + b)^2}{2(a + b)^2}.$$

We note that in order to obtain $f_1(t)$ a knowledge of $\pi(\lambda, t)$ is necessary. Using elementary probability arguments, we find

$$(5.14) \qquad f_1(t) \, dt = \int_{0}^{\infty} \pi(\lambda, t)\lambda \, d\lambda \, dt.$$

Thus we have

$$(5.15) \qquad f_1(t) = p(1, t).$$

The mean number of events that have occurred between 0 and t is given by

$$(5.16) \qquad \overline{n(t)} = \int_{0}^{t} f_1(t) \, dt$$
$$= \frac{a\lambda_0 t}{a + b} + \frac{b\lambda_0}{(a + b)^2} (1 - e^{-(a+b)t}).$$

To obtain the mean square number of events, we must determine $f_2(t_1, t_2)$. In view of the non-Markovian nature of the process, it is convenient to introduce the function $\pi(\lambda_2, t_2|\lambda_1, t_1)$ where $\pi(\lambda_2, t_2|\lambda_1, t_1) \, d\lambda_2$ represents the probability that $\lambda(t_2)$ has a value between λ_2 and $\lambda_2 + d\lambda_2$ at t_2 given that $\lambda(t)$ had a value λ_1 at $t = t_1$ and a value λ_0 initially. Then it is easy to find

$$(5.17) \qquad f_2(t_1, t_2) = \int_{\lambda_1} \int_{\lambda_2} \pi(\lambda_1, t_1) \, d\lambda_1 \lambda_1 \lambda_2 \pi(\lambda_2, t_2|\lambda_1 - b, t_1) \, d\lambda_2$$

$$= \int_{\lambda_1} E\{\lambda(t_2)|\lambda_1 - b, t_1\} \lambda_1 \pi(\lambda_1, t_1) \, d\lambda_1$$

where $E\{\lambda(t_2)|\lambda_1 - b, t_1\}$ represents the conditional moment of λ at t_2 given that λ had a value $\lambda_1 - b$ at t_1. Once we obtain this conditional moment, we can calculate $f_2(t_1, t_2)$ explicitly.

Using arguments similar to those in Section 3, we find that $\pi(\lambda_2, t_2|\lambda_1, t_1)$ satisfies the equation

$$(5.18) \qquad \frac{\partial \pi(\lambda_2, t_2|\lambda_1, t_1)}{\partial t_2} = - a(\lambda_0 - \lambda_2) \frac{\partial \pi(\lambda_2, t_2|\lambda_1, t_1)}{\partial \lambda_2}$$
$$+ (a - \lambda_2)\pi(\lambda_2, t_2|\lambda_1, t_1)$$
$$+ (b + \lambda_2)\pi(\lambda_2 + b, t_2|\lambda_1, t_1)$$

with the initial condition

$$(5.19) \qquad \pi(\lambda_2, t_2|\lambda_1, t_1) = \delta(\lambda_2 - \lambda_1).$$

Equation (5.18) is very similar to (5.8) and hence cannot be solved explicitly. However, we can obtain the conditional moments of λ and those are precisely the quantities that will be needed for the calculation of the correlation functions. On evaluating the conditional moments and using (5.17), we finally obtain

$$(5.20) \quad f_2(t_1, t_2) = p(2, t_1) \, e^{-(a+b)(t_2 - t_1)}$$
$$+ p(1, t_1)\left\{\frac{a\lambda_0}{a + b} [1 - e^{-(a+b)(t_2 - t_1)}] - b \, e^{-(a+b)(t_2 - t_1)}\right\}.$$

An interesting feature that emerges from (5.20) is the existence of the limit of $f_2(t_1, t_2)$ where both t_1 and t_2 tend to infinity, $t_2 - t_1$ remaining a constant τ.

Thus we have

$$(5.21) \qquad \lim f_2(t_1, t_2) = \left(\frac{a\lambda_0}{a+b}\right)^2 - \frac{a\lambda_0 b(2a+b)}{2(a+b)^2} e^{-(a+b)\tau}.$$

Using (5.21) and (5.20), we can calculate the power spectrum of the response (which is nothing but the Fourier transform of $\overline{r(t_1) \, r(t_2)}$) which is a function of $t_2 - t_1$ in the limit.

5.2. FLUCTUATION OF PHOTOELECTRONS*

Intensity correlation experiments to determine the coherence properties in light beams are performed by allowing the light to be incident on a fast photoelectric detector. If we treat the problem semiclassically a stochastic description of the photoelectric counts actuated by the incident radiation field may lead to a probability frequency function $P(n, T)$ governing n, the number of counts in the time interval T under certain conditions to be described later. But when the incident beam possesses certain statistical features which make themselves manifest through correlations in its intensity $I(t)$ at different time intervals, it is rather difficult to arrive at $P(n, T)$ explicitly. However, since such a process can be viewed as a point process defined on the t-space, a description in terms of product densities will be very useful.

We can define the probability that a count occurs in a time interval between t and $t + dt$ to be $\alpha I(t)dt$ where α is the sensitivity of the detector taken to be a constant, while $I(t)$ the intensity of the radiation falling on the detector is given by

$$(5.22) \qquad I(t) = V^*(t) \, V(t)$$

$V(t)$ being the usual analytic signal corresponding to the radiation field. The average number of counts in the time interval $(0, T)$ is given by

$$(5.23) \qquad \bar{n} = \alpha \int_0^T I(t) \, dt.$$

If $f_1(t_1)$, $f_2(t_1, t_2)$ are the product densities of events on the t-axis, then it is easy to see that

$$(5.24) \qquad \begin{aligned} f_1(t) &= \alpha \, E\{I(t)\} \\ f_2(t_1, t_2) &= \alpha^2 \, E\{I(t_1) \, I(t_2)\} \\ f_3(t_1, t_2, t_3) &= \alpha^3 \, E\{I(t_1) \, I(t_2) \, I(t_3)\} \end{aligned}$$

* The treatment outlined in this section is due to Srinivasan and Vasudevan [32].

where the expectation is to be evaluated with the help of the joint probability frequency function of the analytic signals at different times. At first sight, it might appear that the problem is simpler than the space charge limited shot noise. However, the complexity of the present problem lies in the continuous random nature of $I(t)$. In fact, the problem will be intractable if we do not assume Gaussian character of the thermal source combined with some elegant choice of coherence functions.

We shall proceed to determine $P(n, T)$ the probability of the number of photo counts detected in a time interval $(0, T)$. To do this, we first derive an explicit expression for $L(u)$, the product density generating functional as defined by (3.19). It is interesting to note that in the present case the actual correlation functions defined by $g_n(t_1, t_2, \ldots, t_n)$ are identical with the coherence functions $\Gamma_n(t_1, t_2, \ldots, t_n)$ defined by

$$(5.25) \quad \Gamma_n(t_1, t_2, \ldots, t_n) = \overline{[V^\star(t_1)V^\star(t_2) \cdots V^\star(t_n)V(t_1)V(t_2) \cdots V(t_n)]}.$$

If we can make some reasonable assumption for coherence functions, then (3.19) can be used to obtain $P(n, T)$ explicitly. Let us assume that the intensities $I(t)$ are distributed according to the Gaussian Law as in the case of thermal light so that all orders of coherence functions can be expressed in terms of the coherence function of order 2. Thus we may postulate that

$$(5.26) \quad \Gamma_m(t_1, t_2, \ldots, t_m) = (m - 1)!\Gamma_2(t_1 - t_2)\Gamma_2(t_2 - t_3) \cdots$$
$$\Gamma_2(t_m - t_{m-1}).$$

If, in addition, we assume that $\Gamma_2(t_1 - t_2) = \alpha^2 I$, we obtain

$$(5.27) \qquad L([u], T) = [1 - \alpha I \int_0^T u(t)\, dt]^{-1}$$

from which we can identify the generating function of $P(n, T)$ defined by

$$(5.28) \qquad h(z, T) = \Sigma P(n, T)z^n$$

by observing that

$$(5.29) \qquad h(z, T) = L([z - 1], T)$$
$$= [1 + \alpha I T - \alpha I z T]^{-1}.$$

The above equation identifies $P(n, T)$ with the Boson distribution. Our demonstration leading to the Bose-Einstein distribution shows that we can always arrive at $P(n, T)$ starting with the experimentally observed coherence functions In particular, the Bose-Einstein distribution can be confirmed if the coherence functions of a few more orders are available. This is the decided advantage of the product density technique.

5.3. QUEUING PROCESSES

The theory of queuing processes constitutes one of the important branches of stochastic theory and an innumerable number of contributions have been published. The spate of publications continues with unabated vigor and Professor Kendall [4], in one of his valuable surveys presented at the Seventh All Union Conference on the Theory of Probability and Mathematical Statistics, observed that on an average, one paper per week is devoted to the theory of queues. Quite a few monographs have been published on this subject and as we have mentioned in the introductory remarks, we wish to indicate in an explicit manner the use of the product densities in the formulation of the problems.

We shall confine ourselves to single server systems, the methods proposed being applicable to other systems as well. Customers are assumed to arrive at the service counter in a random manner, the time t between two successive arrivals being independently distributed and governed by the probability frequency function (p.f.f.), $f(t)$. The service times are likewise assumed to be independently distributed, $\alpha(t)$ being the corresponding p.f.f. The primary quantities of interest are

  (i) the waiting time $w(t)$ of any fictitious customer arriving at an arbitrary time t
  (ii) the number of customers in the queue at any time t, and
  (iii) the number of customers served during any time interval (T_1, T_2).

Though the service times and the customer inter-arrival times can be considered as renewal processes, the queuing process which is a fusion of these renewal processes has an intricate non-Markovian structure. In view of this difficulty, one of the constituent renewal processes has been assumed to be a simple Markovian process of the Poisson type and a number of interesting results have been obtained in this direction. A general approach due to Kendall [33] deserves special mention. The technique consists in defining a suitable Markov chain and imbedding the queuing process in it. Very recently, the author and his collaborators [34, 35] have imbedded the general queuing process in a suitable renewal process and obtained a number of useful results using the product density technique. In the next few paragraphs, we shall outline this method of approach.

Let the counter be empty at time $t = 0$ when a customer just arrives. Normally, by the time the customer leaves the counter, there would have been further arrivals and the counter remains busy for sometime. If we assume that the counter becomes free at some time $\tau > 0$ and the next arrival occurs between t and $t + dt(t > \tau)$, then we can visualize the interval $(0, t)$ to be a cycle. It is clear that the lengths of such cycles are independently distributed on the basis of the assumptions given above. Thus the epochs of time t_i at which arrivals occur (the counter being free) constitute a renewal point process. This renewal process may be called a "master" renewal process. If $h(t)$ is the p.f.f. governing the duration of such a renewal process, a number of interesting results can be obtained from a knowledge of $h(t)$; $h(t)$ itself can be determined in terms of $\alpha(t)$ and $f(t)$ in special cases (see reference [34]). It may be fruitful to define $h(\tau, t)$ as the joint p.f.f. of the busy period τ and the cycle t in which case results relating to duration of busy and idle periods can be obtained. Once the queuing process is identified with this type of renewal process through $f(t)$, many of the results of renewal theory mentioned in the earlier section become applicable in the present case.

There are other types of product densities that can be introduced. For example, we can ask the question: What is the joint probability that a customer arrives between t and $t + dt$ and completes his service after a time lying in the interval between $(w, w + dw)$? If $F(w, t) \, dt \, dw$ denotes this probability, then $F(w, t)$ is a product density in the product space of t and w. For example, the mean number of customers who have waited for a time longer than w in a period $(0, T)$ is given by

$$\int\limits_{w}^{\infty} dw' \int\limits_{0}^{T} F(w', t') \, dt'.$$

This mean number will be a useful criterion for opening more counters. Work in this direction and other similar problems is in progress and the interested reader is referred to references [34] and [35].

5.4. KINETIC THEORY OF FLUIDS

In kinetic theory, we normally derive macroscopic properties of a system from its microscopic constituents. The system is essentially a blob consisting of a large number of molecules. If $\mathbf{p}$ and $\mathbf{q}$ denote the momentum and position of a molecule, basic quantities like mass, momentum, and local energy (temperature) of a macroscopically small blob of fluid are given by

(5.30) $\qquad M = \Sigma m \, dN(\mathbf{p}, \mathbf{q}; t)$

(5.31) $\qquad P = \Sigma \mathbf{p} \, dN(\mathbf{p}, \mathbf{q}; t)$

(5.32) $\qquad Q = \Sigma \, dN(\mathbf{p}, \mathbf{q}, t)(\mathbf{p} - \bar{\mathbf{P}})^2/2m$

where the summation is over the molecules of the blob and $dN(\mathbf{p}, \mathbf{q}, t)$ represents the number of molecules in an elemental volume of phase-space $d\mathbf{p}\, d\mathbf{q}$. According to the principles of statistical mechanics, the molecules are distributed in phase-space in a random manner. Thus equations (5.30) through (5.32) can be interpreted in a different manner. With each random point in the phase-space (which corresponds to a molecule) we associate a deterministic function and the quantities of interest like mass, momentum, and local energy turn out to be the sum of such suitable functions. Thus the moments of these quantities can be obtained provided we know the product densities explicitly. For instance, the first two moments of P_i are given by

$$(5.33) \qquad \overline{P_i} = \iint p_i f_1(\mathbf{p}, \mathbf{q}; t)\, d\mathbf{p}\, d\mathbf{q}$$

$$(5.34) \qquad \overline{P_i^2} = \iint p_i^2 f_1(\mathbf{p}, \mathbf{q}; t)\, d\mathbf{p}\, d\mathbf{q}$$
$$+ \iiiint p_{1i} p_{2i} f_2(\mathbf{p}_1, \mathbf{q}_1; \mathbf{p}_2, \mathbf{q}_2; t)\, d\mathbf{p}_1\, d\mathbf{q}_1\, d\mathbf{p}_2\, d\mathbf{q}_2.$$

At this stage, if we make the Poisson approximation for the product densities

$$(5.35) \qquad f_2(\mathbf{p}_1, \mathbf{q}_1; \mathbf{p}_2, \mathbf{q}_2; t) = f_1(\mathbf{p}_1, \mathbf{q}_1; t) f_1(\mathbf{p}_2, \mathbf{q}_2; t)$$

equation (5.34) makes the fluctuation theorem of statistical mechanics manifestly clear since the first term on the right-hand side representing the fluctuation about the mean is smaller than the second which is the square of the mean by a factor $1/N$ where N is the number of molecules contained in a macroscopic element of volume. This by itself is an interesting result of far-reaching consequence since we have not made use of any specific form of the first order product density.

There are many interesting and useful results that can be derived from the product density approach presented above and these can be found in references [36] and [37]. There is one feature characteristic of the above type of product densities and distinctly different from that of the branching phenomena. The above product densities obey a hierarchical type of equations. We shall illustrate this by taking the case of a rare gas where the macroscopic properties are determined by collisions of molecules. In the absence of external forces $f_1(\mathbf{p}, \mathbf{q}; t)$ satisfies the equation

$$(5.36) \qquad \frac{\partial f_1(\mathbf{p}, \mathbf{q}; t)}{\partial t} + \frac{p_i}{m} \frac{\partial f_1(\mathbf{p}, \mathbf{q}; t)}{\partial q_i} = \iint [f_2(\mathbf{p}', \mathbf{q}; \mathbf{p}_1', \mathbf{q}_1; t)$$
$$- f_2(\mathbf{p}, \mathbf{q}; \mathbf{p}_1, \mathbf{q}_1; t)] J\, d\mathbf{p}_1\, d\mathbf{q}_1$$

where J depends on a relative momentum of the two colliding molecules, impact parameter, and the azimuthal angle of the plane of collision. Equation

(5.36) is derived using the classic arguments of Boltzmann combined with the probabilistic interpretation of the product densities. Thus we obtain a chain of equations expressing f_2 in terms of f_3, f_3 in terms of f_4, and so on. It is the hierarchical property that characterizes the non-Markovian nature of the process. The kinetic theorists, as we have explained in the previous chapter, were mainly concerned with such hierarchical equations satisfied by the corresponding Janossy densities. The author has proposed a novel approach to the kinetic theory of fluids with the help of the product density functions. This kind of approach is useful in the understanding of a variety of nonequillibrium phenomena such as turbulence and phase transition.

REFERENCES

1. A. Ramakrishnan, *Proc. Camb. Phil. Soc.*, **46** (1950), 595.
2. A. Ramakrishnan, *Proc. Camb. Phil. Soc.*, **49** (1953), 473.
3. S. K. Srinivasan, *J. Math. Phys. Sci.*, **1** (1967), 1.
4. D. G. Kendall, *Theory of Probability and Its Applications*, **9** (1964), 1.
5. M. S. Bartlett, *J. Roy. Statist. Soc.*, B **25** (1963), 264.
6. M. S. Bartlett, *Stochastic Processes*, Cambridge University Press, 1955.
7. D. G. Kendall, *J. Roy. Statist. Soc.*, B **11** (1949), 230.
8. P. I. Kuznestov and R. L. Stratonovich, *Izvestya Akd. Nauk. U.S.S.R.*, Ser. Mat. **20** (1956), 101.
9. A. Ramakrishnan, *Stochastic Processes and Their Applications to Physical Problems*, Ph.D. Thesis, University of Manchester, 1951.
10. P. I. Kuznestov, R. L. Stratonovich, and V. I. Tikhonov, *Non-Linear Transformations of Stochastic Processes*, Pergamon Press, London, 1965.
11. H. J. Bhabha, *Proc. Roy. Soc.* (London), **202 A** (1950), 301.
12. S. K. Srinivasan, *Zastosowania Matematyki*, **6** (1961–62), 209.
13. R. L. Stratonovich, *Topics in the Theory of Random Noise*, Vol. 1, Gordon and Breach, New York, 1963.
14. J. Beutler and O. A. Z. Leneman, *Act. Math.*, **116** (1966), 159.
15. H. Wold, in *Le Calcul des probabilités et ses applications*, C.N.R.S., Paris, 1949.
16. J. E. Moyal, *Act. Math.*, **108** (1962), 1.
17. A. Ramakrishnan and S. K. Srinivasan, *Publication de L'Institut Statistique*, Paris, **5** (1956), 95.
18. A. Ramakrishnan and P. M. Mathews, *Phil. Mag.*, **44** (1953), 1122.
19. D. R. Cox and W. L. Smith, *Biometrika*, **41** (1954), 91.
20. D. R. Cox, *Renewal Theory*, Methuen, London, 1962.
21. J. A. McFadden, *J. Roy. Statist. Soc.*, B **24** (1962), 364.
22. J. A. McFadden and W. Weissblum, *J. Roy. Statist. Soc.*, B **25** (1963), 413.
23. J. A. McFadden, *Sankhya*, **27 A** (1965), 83.
24. M. C. Wang and G. E. Uhlenbeck, *Rev. Mod. Phys.*, **17** (1945), 323.

25. S. K. Srinivasan, *Some Applications of Stochastic Theory to Physical Problems*, M.Sc. Thesis, University of Madras, 1955.
26. P. M. Mathews and S. K. Srinivasan, *Proc. Nat. Inst. Sci.* (India) **22 A** (1956), 369.
27. D. R. Cox, *J. Roy. Statist. Soc.*, **B 17** (1955), 129.
28. L. Takacs, *Act. Math. Acad. Sci.* (Hungary), **5** (1954), 203.
29. E. N. Rowland, *Proc. Camb. Phil. Soc.*, **33** (1937), 344.
30. S. K. Srinivasan, *On a Class of Non-Markovian Processes*—I.I.T. preprint, August, 1962.
31. S. K. Srinivasan, *Nuovo Cimento*, **38** (1965), 978.
32. S. K. Srinivasan and R. Vasudevan, *Nuovo Cimento*, **47** (1967), 185.
33. D. G. Kendall, *J. Roy. Statist. Soc.*, **B 13** (1951), 151.
34. S. K. Srinivasan and R. Subramanian, in *Symposia in Theoretical Physics*, Vol. 9, Plenum Press, New York (1968).
35. S. K. Srinivasan and R. Vasudevan (to be published; 1969).
36. S. K. Srinivasan, *Zeit. fur Physik*, **193** (1966), 394; **197** (1966), 435.
37. S. K. Srinivasan, in A. Ramakrishnan, *Symposia in Theoretical Physics*, Vol. 7, Plenum Press, New York, 1968.

Chapter 4

MULTIPLE PRODUCT DENSITIES AND SEQUENT CORRELATIONS

1. Introduction

In the last two chapters, we have dealt with point processes by the introduction of certain correlation functions which describe the distribution of entities characterized by a continuous parameter. The continuous parameter which we denote by x may refer to the energy of a cosmic ray particle or the age of an individual in a birth and death process or the velocity and position of a fissionable particle. The theory of product densities and cumulant functions introduced in the previous chapters is based on the assumption that the probability that an entity has a value in an infinitesimal interval Δ of parametric space is of the order Δ while the probability of more than one entity having a parametric value within Δ is vanishingly small compared to Δ. Thus it might appear that this method is not suited for the description of processes in which the probability that more than one entity has a parametric value within Δ is of the same order of magnitude as Δ. It is particularly so in the case of population growth where twins and higher multiplets do exist. We shall show in Section 2 that it is possible to extend the theory of product densities to include "multiple points" as well.

In our earlier discussions on point processes, we have not paid much attention to the nature of the continuous parameter x characterizing the points or entities. If x denotes the age of an individual in a birth and death process, the age is a deterministic function in the sense that if the time t_0 of birth is known, the age of the individual can be specified at any later time $t > t_0$, the parameter t standing for time in this case. However, there are numerous physical processes where the time or position in t-axis of the creation of a particle and its then x-value do not determine its subsequent x-value. A concrete example is provided by the energy state x of an electron in the electron-photon cascade. In this case, the multiplicative stochastic process evolves with respect to both t and x. In fact, it makes sense to talk of the parametric value at the point of the creation of the particle. Experimentally, it is convenient to make measurements near the point of production.

53

Thus we can talk of product densities in the product space of x and t. Such product densities have been used in the formulation of cascade theories and will be discussed in Chapter 5 and 6. However, the point which we wish to project at the moment is the possibility of extending product densities to describe correlations at different x and t, t being the parameter with respect to which the process evolves. Such correlations have been called sequent product densities (see Ramakrishnan and Radha [1]) and evolutionary sequent correlations (see Srinivasan et al. [2, 3]. The purpose of this chapter is to present these extensions of the product density technique which includes multiple points and multiple correlations. Sections 2 and 3 will deal with the multiple points. We then introduce the sequent correlations and discuss their relationship to the characteristic functional.

2. Multiple Product Densities—General Formula for the Moments

Let $M(a, b; t)$ be the random variable representing the number of entities with parametric values between a and b; t here represents the parameter with respect to which the process evolves. We shall assume that the maximum number of entities having parametric values in the infinitesimal range dx is n. In counting the number of entities, an i-tuplet will be counted as i entities since the members of a multiplet evolve individually with respect to the parameter. Our starting point is the density function introduced by Janossy [4] suitably modified to accommodate multiplets.

We shall consider

$$J_{m_1 m_2 \ldots m_n} (x_1^1, x_2^1, \ldots, x_{m_1}^1, \ldots x_1^n, x_2^n, \ldots, x_{m_n}^n; t)$$

where

$$J_{m_1 m_2 \ldots m_n} \prod_{ij_i} dx_{j_i}^i \, (j_i = 1, 2, \ldots m_i, i = 1, 2, \ldots n)$$

represents the joint probability that there exist m_i i-tuples and that a typical i-tuple has a parametric value between

$$x_{j_i}^i \quad \text{and} \quad x_{j_i}^i + dx_{j_i}^i.$$

To obtain information regarding the moments of the total number of entities having some parametric value in the range (a, b) we consider an assembly of entities $(m_1, m_2, \ldots, m_n)$, where m_i is the number if i-tuples, the i-tuples having parametric values in the intervals

$$(x_1^i, x_1^i + dx_1^i), (x_2^i, x_2^i + dx_2^i) \ldots, (x_{m_i}^i, x_{m_i}^i + dx_{m_i}^i).$$

Then the random variable $M(a, b; t)$ takes the value

$$\sum_{i=1}^{n} \sum_{j_i=1}^{m_i} iH(x_{j_i}^i - a)H(b - x_{j_i}^i)$$

with probability

$$J_{m_1 m_2 \ldots m_n} \prod_{ij_i} dx_{j_i}^i,$$

$H(x)$ being the Heaviside unit function. The mean of the total number of entities having parametric values in the range (a, b) is given by

$$E\{M(a, b; t)\}$$

$$= \sum_{i=1}^{n} \sum_{m_i} \sum_{j_i=1}^{m_i} \int dx_1^1 \int dx_2^1 \cdots \int dx_{m_1}^1 \cdots \int dx_1^n \int dx_2^n \cdots \int dx_{m_n}^n$$

$$J_{m_1 m_2 \ldots m_n} \frac{iH(x_{j_i}^i - a)H(b - x_{j_i}^i)}{m_1! m_2! \cdots (m_i - 1)! \cdots m_n!}$$

$$= \sum_{i=1}^{n} i \int_a^b dx \left[\sum_{m_1=0}^{\infty} \cdots \sum_{m_i=1}^{\infty} \cdots \sum_{m_n=0}^{\infty} \frac{1}{m_1! m_2! \cdots (m_i - 1)! \cdots m_n!} \right.$$

$$\int_0^\infty dx_1^1 \int_0^\infty dx_2^1 \int_0^\infty dx_{m_1}^1 \cdots \int_0^\infty dx_1^i \int_0^\infty dx_2^i \cdots \int_0^\infty dx_{m_i-1}^i$$

$$\int_0^\infty dx_1^{i+1} \int_0^\infty dx_2^{i+1} \cdots \int_0^\infty dx_{m_n}^n$$

$$J_{m_1 m_2 \ldots m_n} (x_1^1, x_2^1, \ldots, x_{m_1}^1, \ldots, x_1^i, x_2^i, \ldots, x_{m_i-1}^i, x_1^{i+1},$$

$$\left. x_2^{i+1}, \ldots, x_1^n, x_2^n, \ldots, x_{m_n}^n) \right].$$

We immediately recognize that the quantity within the square bracket denotes the probability that there is an i-tuple having a parametric value between x and $x + dx$ irrespective of the number elsewhere. Writing

$$(2.2) \qquad E\{M(a, b; t)\} = \sum_{i=1}^{n} i \int_a^b f_1^i(x, t) \, dx = \int_a^b f_1(x, t) \, dx$$

we can identify $f_1(x, t)$ in the notation of Ramakrishnan as the product density of degree one. Thus we can call $f_1^i(x, t)$ the product density of degree one of i-tuples. It is well recognized that it is easy to derive differential equations for $f_1^i(x, t)$ in any particular process.

We next proceed to obtain the ν-th moment to identify the product density of degree ν and the manner in which contributions from the product densities of lower order arise.

Writing

$$(2.3) \qquad\qquad g(x_{j_i}^i) = H(x_{j_i}^i - a)H(b - x_{j_i}^i)$$

we observe that

$$(2.4) \qquad [\sum_{i=1}^{n} i \sum_{j_i=1}^{m_i} g(x_{j_i}^i)]^{\nu}$$

$$= \sum_{s=1}^{\nu} \sum_{i_1=1}^{n} \sum_{i_2=1}^{n} \cdots \sum_{i_s=1}^{n} \underset{h_1+h_2+\cdots+h_s=v}{\sum_{h_1} \sum_{h_2} \cdots \sum_{h_s}} \sum_{l_1=1}^{h_1} \sum_{l_2=1}^{h_2} \cdots \sum_{l_s=1}^{h_s} \frac{\nu!}{h_1! h_2! \cdots h_s!}$$

$$\frac{m_{i_1}! \, m_{i_2}! \cdots m_{i_s}!}{l_1! \, l_2! \cdots l_s!} \frac{C_{l_1}^{h_1} C_{l_2}^{h_2} \cdots C_{l_s}^{h_s}}{(m_{i_1} - l_1)! \, (m_{i_2} - l_2)! \cdots (m_{i_s} - l_s)!}$$

$$i_1^{h_1} i_2^{h_2} \cdots i_s^{h_s} \, g(x_1^{i_1}) \, g(x_2^{i_1}) \cdots g(x_{l_1}^{i_1}) \cdots g(x_1^{i_s}) \, g(x_2^{i_s}) \cdots g(x_{l_s}^{i_s})$$

where C_s^{ν} denotes the number of ways in which $(\nu - s)$-fold degeneracy can arise in a product of ν terms. Using (2.4) and observing that

$$\sum \frac{1}{m_1! \, m_2! \cdots (m_{i_1} - l_1)! \cdots (m_{i_s} - l_s)! \cdots m_n!}$$

$$\int dx_1^1 \int dx_1^2 \cdots \int dx_{m_1}^1 \cdots \int dx_1^n \int dx_2^n \cdots \int dx_{m_n}^n \, J_{m_1 m_2 \ldots m_n}$$

$$g(x_1^{i_1}) \, g(x_1^{i_1}) \cdots g(x_{l_1}^{i_1}) \cdots g(x_1^{i_s}) \, g(x_2^{i_s}) \cdots g(x_{l_s}^{i_s})$$

could be written as

$$\int_a^b dx_1 \int_a^b dx_2 \cdots \int_a^b dx_{k_s} \, f_{k_s}^{l_1, l_2, \ldots, l_s} (x_1, x_2, \ldots, x_{k_s}; t)$$

we can express the ν-th moment of the variable $M(a, b; t)$ as

(2.5) $\quad E\{[M(a, b; t)]\}^\nu$

$$
= \sum_{\substack{s=1 \\ i_1 \neq i_2 \neq}}^{\nu} \sum_{\substack{i_1=1}}^{n} \sum_{\substack{i_2=1}}^{n} \cdots \sum_{\substack{i_s=1 \\ \cdots \neq i_s}}^{n} \sum_{\substack{h_1=1 \\ h_1 + h_2 +}}^{\nu} \sum_{\substack{h_2=1 \\ \cdots + h_s = v}}^{\nu} \cdots \sum_{\substack{h_s=1}}^{\nu}
$$

$$
\sum_{l_1=1}^{h_1} \sum_{l_2=1}^{h_2} \cdots \sum_{l_s=1}^{h_s} \frac{\nu! \, i_1^{h_1} i_2^{h_2} \cdots i_s^{h_s}}{h_1! h_2! \cdots h_s! \, l_1! \, l_2! \cdots l_s!} \, C_{l_1}^{h_1} C_{l_2}^{h_2} \cdots C_{l_s}^{h_s}
$$

$$
\int_a^b dx_1 \int_a^b dx_2 \cdots \int_a^b dx_k f_{k_s}^{l_1, l_2, \cdots, l_s} (x_1, x_2, \ldots, x_{k_s}; t)
$$

where $f_k^{l_1, l_2, \cdots l_s} (x_1, x_2, \ldots, x_{k_s}; t) \, dx_1 \, dx_2 \cdots dx_{k_s}$ represents the joint probability that there are l_1 i_1-tuples, l_2 i_2-tuples $\cdots$, l_s i_s-tuples, the parametric values of these entities lying respectively in the ranges

$$(x_1, x_1 + dx_1), (x_2, x_2 + dx_2) \cdots, (x_{k_s}, x_{k_s} + dx_{k_s})$$

irrespective of the distribution of the entities elsewhere.

For any process, we can deal with

$$f_{k_s}^{l_1, l_2, \cdots, l_s} (x_1, x_2, \ldots, x_{k_s}; t)$$

and obtain differential or integral equations by studying the evolution of the process with respect to t or x or both as the case may be.

3. Multiple Product Densities—An Alternative Approach

The theory of multiple product densities presented in the previous section does not bring out the properties of the random variable $dN(x, t)$ or its correlations. An attempt in this direction was made sometime back by Ramakrishnan and Srinivasan (1958) who dealt with the problem of age distribution in population growth involving twins and higher multiplets. The product density technique has been extended by treating the process as a vector process, the different members of the multiplet forming the components of the vector. The calculations are confined to the second moment of the number distribution. The general problem is not trivial since the quantity of interest is the total number of entities irrespective of the order of the multiplet to which an entity belongs. In this context, the moment formula (2.5) is useful even though the product densities are introduced in an indirect manner and presuppose a knowledge of the properties of Janossy functions. We shall here present a direct method which projects the random variable $dN(x, t)$ to the fore (see Ramakrishnan *et al.* [5]).

Exactly as in the case of product densities, we introduce $N_i(x, t)$ as the stochastic variable representing the number of i-tuple entities having parametric values less than or equal to x. We assume as before that the probability that there is an i-tuple with parametric value in $(x, x + \delta x)$ is proportional to δx the probability of more than one i-tuple with parametric value in the range $(x, x + \delta x)$ being of a smaller order of magnitude than δx. Thus it is possible to define the function $f_1^i(x; t)$ $(i = 1, 2, \ldots, m)$* such that

$$(3.1) \qquad\qquad f_1^i(x; t)\delta x = E\{dN_i(x; t)\}.$$

If $P_i(n)$ is the probability that n i-tuples occur in δx, then

$$(3.2) \qquad P_i(1) = E\{dN_i(x, t)\} + o(\delta x)$$
$$= f_1^i(x, t)\delta x + o(\delta x)$$

$$P_i(0) = 1 - f_1^i(x, t)\delta x + o(\delta x)$$

$$P_i(n) = o(\delta x) \qquad\qquad\qquad n > 1.$$

In addition we postulate that $P_{ij}(1, 1)$, the probability of finding an i-tuple and a j-tuple in δx, is $o(\delta x)$. The random variables $dN_i(x, t)$ enjoy the same properties as the variable $dN(x, t)$ introduced in Section 2 of Chapter 3. However, we notice that the random variable that is of interest is $dM(x, t)$ given by

$$(3.3) \qquad\qquad dM(x, t) = \sum_i i\, dN_i(x, t)$$

The mean number of entities having parametric values in the interval (a, b) is given by

$$(3.4) \qquad E\{M(b; t) - M(a; t)\} = \sum_i i \int_a^b E\{dN_i(x, t)\} = \Sigma i \int_a^b f_1^i(x, t)\, dx$$

where $f_1^i(x, t)$ is the product density of degree one of i-tuples. Thus the expression for the mean does not offer any special difficulty since we are calculating the expectation value of a linear function. However, if we wish to calculate the higher moments of the variable $M(b, t) - M(a, t)$ we have to encounter certain new features in the combinatorial analysis of the problem

* As before, we shall assume that m is the highest order of multiplicity.

arising from (i) m the highest order of multiplicity of the entities and (ii) ν the order of the moment that is sought. However, the problem can be solved easily for any given m and especially when m and ν are small. To evaluate the ν-th moment of $M(b, t) - M(a, t)$ in the general case, we must fix our attention on the number of ways in which the degeneracies arise from the overlapping of dx_i's. In the present case, the procedure is more complicated than in the case of simple product densities and it is necessary to break C_h^ν into "complexions" (as has been done in Section 3 of Chapter 3). Denoting the number of ways of obtaining a typical complexion belonging to (h, ν) as $C_h^\nu(\alpha)$, we obtain

(3.5) $\quad E\{[M(b, t) - M(a, t)]^\nu\}$

$$= \sum_k \sum_\alpha \sum_{i_1=1}^m \sum_{i_2=1}^m \cdots \sum_{i_h=1}^m C_h^\nu(\alpha)\, i_1^{l_1}\, i_2^{l_2} \cdots i_h^{l_h}$$

$$\int_a^b dx_1 \int_a^b dx_2 \cdots \int_a^b dx_h f_{i_1\, i_2\, \cdots\, i_h}(x_1, x_2, \ldots, x_h; t)$$

where the summation over h is to be properly interpreted since the complexions will arise only if $h > 1$ or $< m - 1$; $f_{i_1\, i_2\, \cdots\, i_h}$ is the product density of degree h of an i_1-tuple, an i_2-tuple, $\ldots$ and an i_h-tuple, the integers $i_1, i_2, \ldots, i_h$ being not necessarily distinct.

It is interesting to compare (3.5) with (2.5) which contains

$$f_{k(s)}^{l_1, l_2, \cdots l_s}(x_1, x_2, \ldots, x_{k(s)}, t)$$

the product density of degree $k(s) = l_1 + l_2 + \cdots + l_s$ composed of l_1 i_1-tuples l_2 i_2-tuples $\cdots$ and l_s i_s-tuples. If it were possible to classify the product densities occurring in (3.5) according to the different number of i-tuples, a relation between the number of complexions $C_h^\nu(\alpha)$ and the coefficients C_l^h can be established and such a relationship may not be trivial even if we consider a specialized case of a given set of entities distributed over the interval $[a, b]$.

The multiple point product densities have been used by Ramakrishnan and Srinivasan [6] in the solution of the fluctuation of the population size when birth rates are assumed to be age specific. K. S. S. Iyer [7] has used the multiple densities for the study of certain models in the theory of carcinogenesis. Ramakrishnan, Vasudevan, and Rajogopal [8] have discussed the problem of neutron transport with fission with the help of multiple product densities. For other examples illustrating the use of multiple point processes, the reader is referred to the original paper of Srinivasan [9].

4. Other Extensions of the Product Density Technique

In our earlier discussion on point processes and product densities, the parameter t was not given any importance. The point of interest is that if we take t to be a one-dimensional parameter (for example time), the correlation functions introduced in Chapter 3 can be called *instant product densities* for the obvious reason that the product densities define the correlation at the particular t. Such a viewpoint enables us to conceive of sequent product densities expressing the correlation of the random variable $dN(x, t)$ at different t's in addition to the x's. This simple idea can be extended a little. To be specific, let us consider a cascade process. If we take for x, the energy of the particle and t the depth traversed by the particle, we can think of the energy x_1, at the point of production t_1* of the particle and the energy x_2 at a later point t_2. Thus we can study the correlation between the "primitive" energy of the particle and its energy at any other depth. This, in turn, leads to yet another type of correlation function connecting the primitive energy of a particle at a certain depth and the energy of a particle at a later depth. We now proceed to a systematic discussion of these correlations in detail.

4.1. SEQUENT PRODUCT DENSITIES

The concept of sequent product density was first introduced by Ramakrishnan and Radha (1960) who distinguished between "instant" and "sequent" correlations in point processes. The instant correlation relates to the study of correlations of the random variables corresponding to the same value of t, while the sequent correlation relates to that between the random variables corresponding to different values of t. Though the arguments used by Ramakrishnan and Radha depend heavily on t being the time parameter, the results are applicable to any ordered parameter. In fact, if $N(x, t)$, in the notation of Chapter 3, is the stochastic variable representing the number of entities having a parametric value $X < x$, then the sequent product densities can be obtained by considering the expectation value of the product $dN(x_1, t_1)$ $dN(x_2, t_2) \cdots dN(x_m, t_m)$. The sequent product densities contain more information than the instant product densities in that they partly explain the dependence of $N(x_i, t_i)$ on $N(x_j, t_j)(t_i < t_j)$. Thus the second order sequent density is always expressible only in terms of second order instant density.

We shall illustrate this by considering the nucleon cascade. Taking only

* Since the production can be located in an interval $(t, t + dt)$ the correlation function will be a density function in t as well.

one type of particle, it is easy to see that the n-th order sequent density corresponding to different thicknesses is connected to all the instant densities of order m less than or equal to n. We first establish the connection for the second order. If $F_2(E_1, E_2; t_1, t_2|E_0)$ is defined by

$$E\{dN(E_1, t_1)\, dN(E_2, t_2)\} = F_2(E_1, E_2; t_1, t_2|E_0)\, dE_1\, dE_2$$

then for $t_2 > t_1$, we have

$$(4.1) \quad F_2(E_1, E_2; t_1, t_2|E_0) = \int_{E_2}^{\infty} f_2(E_1, E_2'|E_0; t_1) f_1(E_2|E_2'; t_2 - t_1)\, dE_2'$$

$$+ f_1(E_1|E_0; t_1) f_1(E_2|E_1; t_2 - t_1)$$

where the parameter E_0 merely indicates that at $t = 0$ there was a primary of energy E_0. This equation is obtained by the following argument. The particle of energy E_2 at t_2 must belong to a cascade generated either by a particle of energy E_2' at t_1, or by the particle of energy $E_1(>E_2)$ at t_1. The two terms in (4.1) refer to these two possibilities. It is to be noted that F_2 is not symmetric in t_1 and t_2 as is usual in an evolutionary Markovian process.

Defining $N(E, t_1)$ and $N(E, t_2)$ as the random variables representing the number of particles above energy E at t_1 and t_2 respectively, the expectation value of their product is given by

$$(4.2) \quad E\{N(E, t_1)\, N(E, t_2)\} = \int_{E}^{E_0} \int_{E}^{E_0} F_2(E_1, E_2; t_1, t_2)\, dE_1\, dE_2$$

The variation of $E\{dN(E_1, t_1)\, dN(E_2, t_2)\}$ as t_1 varies from 0 to t_2 is not only of considerable mathematical interest but also of physical interest. For, if we assume the entire cascade to be generated by a single particle of energy E_0, the second-order density at $t = 0$ is zero and $f_1(E_1, 0) = \delta(E_0 - E_1)$. Hence, when t_1 approaches 0 the first term of (4.1) vanishes and F_2 reduces to

$$(4.3) \quad F_2(E_1, E_2; 0, t_2) = \delta(E_1 - E_0) f_1(E_2|E_1; t_2).$$

As t_1 tends to t_2, $F_2(E_1, E_2; t_1, t_2)$ does not reduce just to $f_2(E_1, E_2; t_2)$ but to

$$(4.4) \quad F_2(E_1, E_2; t_2, t_2) = f_2(E_1, E_2; t_2) + f_1(E_1, t_2)\delta(E_2 - E_1).$$

Hence $E\{N(E; 0)N(E; t_2)\}$ in the two limiting cases are given by

$$(4.5) \quad E\{N(E; 0)N(E; t_2)\} = \int_{E}^{E_0} f_1(E_2|E_0; t_2)\, dE_2$$

and

$$(4.6) \qquad E\{[N(E, t_2)]^2\} = \int_E^{E_0} \int_E^{E_0} f_2(E_1, E_2; t_2) \, dE_1 \, dE_2$$

$$+ \int_E^{E_0} f_1(E_1; t_2) \, dE_1.$$

In the case of instant product densities, the r-th moment is connected to all moments of order less than r in view of the degeneracies that occur in the density of degree r when the energy variables become equal. It turns out that this is implied in the definition of sequent product density and hence the limiting process $t_1 \to t_2$ in (4.4) gives the mean square number directly and involves the first moment also according to (4.6).

Finally, we wish to observe that the sequent product densities of various orders can be generated by the characteristic functional $\Psi([\theta], T)$ defined by

$$(4.7) \qquad \Psi([\theta], T) = E\{\exp i \int_x \int_0^T dN(x, t) \, \theta(x, t) \, dt\}$$

the functional derivatives of Ψ with respect to θ yielding the desired sequent densities. In order to obtain the sequent densities explicitly, it is necessary to study the variation of Ψ with respect to the initial spectrum of the cascade process. For a cascade generated by a single primary of specified energy, it is possible to obtain differential equations by a method due to Bellman *et al.* [10] which is well known under the name invariant imbedding technique and this will be discussed in Chapter 6 when we deal with the electromagnetic cascades.

4.2. EVOLUTIONARY SEQUENT CORRELATION FUNCTIONS

There is yet another type of product density introduced by Srinivasan and Iyer [2]. We shall call it the evolutionary sequent correlation for the simple reason that it carries relevant information about the evolution of the parametric values of entities. To appreciate such sequent correlations, we introduce the notion of primitive value of an entity. This is defined as the parametric value at its point of inception. We can consider the random variables the primitive value of each of which is greater than or equal to x_1 at $t = t_2$. Then we can define an evolutionary sequent correlation density by considering

$d\mathcal{M}(x_1, t_1; x_2, t_2)$ which denotes the stochastic variable representing the number of entities that are produced between t_1 and $t_1 + dt_1$, the primitive value of each of which is between x_1 and $x_1 + dx_1$, the parametric value of these entities lying between x_2 and $x_2 + dx_2$ at $t = t_2$. We shall reserve the symbol $\mathcal{F}$ to denote such a product density. If $P(n)$ is the probability that the random variable $d\mathcal{M}$ takes the value n, it is reasonable to assume that $P(1)$ is of the order $\delta\Omega$ while $P(n)$ for $n \geqslant 2$ is of a smaller order of magnitude as compared to $\delta\Omega$, $\delta\Omega$ being an infinitesimal element in the space in which the density functions are defined.*

Thus, we can define the sequent correlation density of degree one by

(4.8)
$$P(1) = f_1(x_1, t_1; x_2, t_2)\delta\Omega + o(\delta\Omega)$$
$$= \mathscr{E}\{d\mathcal{M}(x_1, t_1; x_2, t_2)\}$$
$$P(0) = 1 - f_1(x_1, t_1; x_2, t_2)\delta\Omega + o(\delta\Omega)$$
$$P(n) = o(\delta\Omega) \qquad\qquad n > 1.$$

Higher order sequent correlation densities are defined in a similar manner. Correlation density of degree n is defined by

(4.9) $E\{d\mathcal{M}(x_1, t_1,; x_1', t)\, d\mathcal{M}(x_2, t_2; x_2', t) \cdots d\mathcal{M}(x_n, t_n; x_n', t)\}$

$$= F_n(x_1, t_1, x_2, t_2, \ldots, x_n, t_n; x_1', x_2', \ldots, x_n', t)\delta\Omega_1\, \delta\Omega_2 \cdots \delta\Omega_n$$

provided the $\delta\Omega_i$ values are disjoint. If all the $\delta\Omega_i$ are not disjoint, then a degeneracy as in the case of usual product densities occurs. We shall not discuss this any further, since all the steps leading to the moment formula of Section 2 of Chapter 3 are applicable in the present case provided dx is replaced by $\delta\Omega$.

4.3. MIXED SEQUENT CORRELATIONS

If, however, we are interested in the sequent product densities of particles that are created at a certain t_1 and particles that are found at t_2, it is clear that we have to deal with the products of the type $dM(x_1, t_1)\, dN(x_2, t_2)$ where $dM(x_1, t_1)$ defined over the product space of x_1 and t_1 denotes the number of particles produced between t_1 and $t_1 + dt_1$, $dN(x_2, t_2)$ being, of course, defined as usual over the x_2 space. Thus we can generalize and define the sequent correlation by

* Ω is the product space of x_1, x_2, and t_1.

$$(4.10) \quad P_{n,m}(x_1, t_1, x_2, t_2, \ldots x_n, t_n; x_1', x_2', \ldots x_m', t)\, d\Omega_{n,m}$$

$$= E\{dM(x_1, t_1)\, dM(x_2, t_2) \cdots dM(x_n, t_n)\, dN(x_1', t)\, dN(x_2', t) \cdots$$

$$dN(x_m', t)\}$$

$$(x_i, t_i) \neq (x_j, t_j) \quad x_i' \neq x_j'$$

where $d\Omega_{n,m}$ is an element of the product space

$$\Omega_1 \times \Omega_2 \times \cdots \Omega_n \times x_1' \times x_2' \times \cdots x_m'$$

of $(2n + m)$ dimensions.

It is worthwhile to investigate the possibility of defining a suitable characteristic functional defined over the product space of x and t. Such characteristic functionals have not been defined earlier for the simple reason that the evolution of the process with respect to t cannot be discussed. However, if we regard the characteristic functional as a functional of the range of integration $(0, T)$, then the behavior of the characteristic functional can be discussed by the method of invariant imbedding technique. We shall consider the particular case of electromagnetic cascades and illustrate the method of obtaining correlations of the type (4.10) from the characteristic functional as defined by (4.7).

As is customary, we shall assume that the probability, per unit thickness of matter,

(i) that an electron of energy E radiates a quantum and drops to an energy lying between E' and $E' + dE'$ is $R_1(E'|E)\, dE'$ and

(ii) that a photon of energy E creates a pair of electrons one of which has an energy between E' and $E' + dE'$ is $R_2(E'|E)\, dE'$.

We shall assume only these two processes in the development of the cascade and neglect the lateral spread of showers so that the stochastic process governing the development of the shower progresses with respect to the parameter t denoting the thickness of matter. This is an example of a point process defined over a Cartesian product of E and t. However, the appropriate random variables are $dN_1(E, t)$ and $dN_2(E, t)$ denoting respectively the number of electrons and photons that are found at t with energies in the interval $(E, E + dE)$ by a primary of j-th type of energy $E_0(j = 1, 2$ denote respectively an electron and a photon primary). Accordingly, we can define the two characteristic functionals by

$$(4.11) \quad \Gamma_j([\theta_1], [\theta_2], T, E_0) = E\{\exp i \int_0^{E_0} \int_0^T [dN_1(E, t)\theta_1(E, t)$$

$$+ dN_2(E, t)\theta_2(E, t)]\, dt\}$$

where the suffix j denotes the shower excited by a primary of the j-th type.

If we now define $\Lambda_j([\chi_1], [\chi_2], T, E_0)$ as the characteristic functional of electrons and photons produced, then we have the fundamental relation

(4.12) $\qquad \Lambda_j([\chi_1], [\chi_2], T, E_0) = \Gamma_j([\theta_1], [\theta_2], T, E_0)$

provided θ_i is chosen to be the linear functional of χ_{3-i} given by

(4.13) $\qquad \theta_i(E, t) = \int_0^{E_0} \chi_{3-i}(E', t) R_i(E'|E)\, dE'.$

The mixed sequent correlation involving $dM(E, t)$ and $dN(E', t)$ can be obtained by taking an appropriate functional derivative of Γ_j. Thus we have

(4.14) $\qquad \dfrac{\delta^2 \Gamma_j}{\delta\theta_m(E_1, t_1)\delta\chi_m(E_2, t_2)}\bigg|_{\substack{\chi_1=0 \\ \chi_2=0}} dE_2\, dt_2\, dE_1 = E\{dN_m(E_1, t_1)\, dM_m(E_2, t_2)\}$

$\qquad\qquad\qquad\qquad\qquad\qquad\qquad\qquad\qquad\qquad\qquad (m = 1, 2).$

In a similar manner, higher order correlations can be obtained. Explicit Mellin transform solutions for these correlations can be obtained and these will be taken up in their proper context in Chapter 6.

The sequent product densities that can be generated from (4.7) are useful in dealing with two-time correlations in a plasma. Apart from this, these functions may provide an adequate basis for a description of many component plasma processes which have many features similar to the cascade processes.

REFERENCES

1. A. Ramakrishnan and T. K. Radha, *Proc. Camb. Phil. Soc.*, **57** (1961), 843.
2. S. K. Srinivasan and K. S. S. Iyer, *Nuovo Cimento*, **33** (1964), 273.
3. S. K. Srinivasan, N. V. Koteswara Rao, and R. Vasudevan, *Nuovo Cimento*, (1968; to be published).
4. L. Janossy, *Proc. Roy. Irish Acad. Sci.*, **53** A (1950), 181.
5. A. Ramakrishnan, S. K. Srinivasan, and R. Vasudevan, *J. Math. Phys. Sci.*, **1** (1967), 275.
6. A. Ramakrishnan and S. K. Srinivasan, *Bull. Math. Biophys.*, **20** (1958), 288.
7. K. S. S. Iyer, *Stochastic Point Processes and Their Applications to Physical Problems*, Ph.D. Thesis, Indian Institute of Technology, Madras (1964).
8. A. Ramakrishnan, R. Vasudevan, and P. Rajagopal, *J. Math. Analys. Applcns.*, **1** (1960), 145.
9. S. K. Srinivasan, *Zastosowania Matematyki*, **6** (1961–62), 209.
10. R. E. Bellman, R. Kalaba, and G. M. Wing, *J. Math. Phys.*, **1** (1960), 280.
11. D. G. Kendall, *J. Roy. Statist. Soc.*, **B 11** (1949), 230.

Chapter 5

ELECTROMAGNETIC CASCADES—
MATHEMATICAL TECHNIQUES

1. Introduction

When a charged particle traverses matter, it interacts with the nuclei of the matter and consequently loses part of its energy by emitting a photon. The emitted photon, if sufficiently energetic, in turn, interacts with matter and annihilates itself into an electron-positron pair. This process gets repeated and starting from a single electron of a certain energy, it is easy to see that the process gives rise to a shower of electrons and photons. Such a shower of particles generated by the above process is called an electromagnetic cascade, since the very mode of formation of the shower is due to the electromagnetic nature of the interaction of matter with radiation. Such cascades were first discovered in the experimental study of cosmic rays* the primary component of which consists of highly energetic protons. The primary proptons in their passage through upper layers of the atmosphere produce further protons, neutrons (by knocking them out of the nuclei), and other particles called pions. Some of the pions so produced, by their characteristic property decay spontaneously into a pair of photons. These photons along with other possible photons that can be produced directly by the primary protons give rise to an electromagnetic cascade.

An electromagnetic cascade develops predominantly by the process of pair creation by a photon and radiation of a photon by an electron (known as bremsstrahlung), these fundamental processes being governed by certain statistical laws derivable from the fundamental principles of quantum electrodynamics. Once the statistical laws governing pair creation and bremsstrahlung are given, the development of the cascade falls within the realms of stochastic theory. This was first realized simultaneously and independently by Bhabha and Heitler [3] and Carlson and Oppenheimer [4] who arrived

---

* For a detailed account of cosmic rays in all their aspects, the reader is referred to Janossy [1]. The recent monograph of Ramakrishnan [2] reports the state of knowledge to about 1960.

66

at the basic equations of the cascade describing the mean behavior. Subsequently, a large number of papers on this subject have been published and especially in the early 1950's, a substantial part of each issue of leading journals such as *Physical Review* were devoted to some aspect or other of the shower theory (see, for example, reference [5]). Of course, this was partly due to the then accepted view that a deep and comprehensive study of all the aspects of cosmic ray showers might unravel the interaction of the various elementary particles. With the invention of energetic accelerators capable of producing many of the cosmic ray phenomena within the laboratory, the limitations of shower theory particularly in its being the guideline for the understanding of the interaction of the elements has been very well realized (see, for example, reference [2], p. 425). However, the theory, in the opinion of the author (see also Harris [6], p. 165) may be considered to be of basic importance since a knowledge of the complete characteristics of showers is not only essential for the construction of a complete cascade theory but important in application as well. In fact, the data on the electron number fluctuations at different depths of shower development are essential to solve the important methodological problem of efficiency of γ-quantum and electron detectors. Information on the total energy released by shower particles or on the fluctuations of the intensity of Cerenkov radiation emitted by shower particles is extremely useful in the design of total-absorption spectrometers and shower spark detectors which are currently used in accelerator experiments (see, for example, reference [7]).

The stochastic nature of the problem attracted the attention of a number of theoretical physicists and soon concerted attempts were made in the direction of estimating the fluctuations of the showers about the mean size. It was in the course of such attempts, some of the techniques of point processes described in the last three chapters were developed. In the case of electron photon showers, the earliest estimation of the number fluctuation about the mean is that of Scott and Uhlenbeck [8] and this was further refined by Bhabha and Ramakrishnan [9] and Janossy and Messel [10] by introducing entirely different mathematical techniques. An interesting account of the various methods and results is given in the monograph of Bharucha-Reid [11]. We wish to confine our attention mainly to the later developments and certain aspects of analytical and computational methods that have not been covered by Bharucha-Reid.

2. Formulation of the Problem

Throughout this chapter, we shall ignore the lateral spread of the shower and treat the problem as one-dimensional so far as the spatial distribution of the particles is concerned. The stochastic problem of electromagnetic cascades can be stated as follows:

Given the initial energy spectrum of photons and electrons at thickness $t = 0$ and that

(i) an electron of energy E, while traversing a small thickness, radiates a photon and drops to an energy between E' and $E' + dE'$ with a probability $R_1(E'|E)\,dE'$

(ii) a photon of energy E while traversing a small thickness decays into a pair of electrons* one of which has an energy between E' and $E' + dE'$ with a probability $R_2(E'|E)\,dE'$, and

(iii) an electron with energy E loses energy deterministically of magnitude $\beta(E)\Delta$ in its passage through a thickness Δ,

we wish to determine the probability distribution of the number of photons and electrons at any thickness t.

We have already mentioned that the charged particles interact with the nucleus. The distance from the nucleus due to which radiation phenomenon occurs plays an important role in the calculation of the cross-section. It turns out that the radiation processes of electrons take place at distances that are large compared to the nuclear radius so that the nucleus can be regarded as a point charge. However, the electrons in the outer shells of the constituent atom screen the electric field of the nucleus. When "complete" screening is assumed, R_1 and R_2 are given by

$$(2.1) \qquad R_1(E'|E) = \left\{ \frac{E - E'}{E} - \left(\frac{4}{3} + \alpha \right)\left(1 - \frac{E}{E - E'} \right) \right\} \frac{1}{E}$$

$$(2.2) \qquad R_2(E'|E) = \left\{ 1 - \left(\frac{4}{3} + \alpha \right)\left(\frac{E'}{E} - \frac{E'^2}{E^2} \right) \right\} \frac{1}{E}$$

where the cross-sections are given per unit thickness of matter, the thickness itself being measured in a convenient unit called radiation length given by

$$(2.3) \qquad \frac{1}{X_0} = 4\alpha_0 \frac{N}{A} Z^2 r_e^2 \ln (183 Z^{-1/3})$$

where α_0 is the fine structure constant, r_e the classical radius of the electron, and Z the charge of the nucleus. If we neglect process (iii), we shall call it Approximation A. If we include it but assume β to be a constant, we shall call it Approximation B.

* We shall not distinguish between electrons and positrons.

3. Janossy's G-Function Method

As stated in the previous section, our object is to obtain the probability frequency function governing the number of electrons and photons in a shower in which the initial energy spectrum of electrons and photons is specified. The problem is indeed a difficult one for general conditions. In fact, the difficulty lies in our inability to specify the initial spectrum precisely. If, for example, following Janossy [12] we define the function $\pi(n_1, E_1, m_1, E_2, t)$ to denote the probability that there are n_1 electrons above an energy E_1 and m_1 photons above an energy E_2 at t, then the value of the function at $t = 0$ does not determine the value of the function at any $t > 0$. This is due to the fact that the basic stochastic process defined by the random variables n_1 and m_1 is non-Markovian in its character since the information that there are n_1 electrons each with an energy greater than E_1 and m_1 photons each with an energy greater than E_2 at t is insufficient to predict their behavior in the interval $(t, t + \Delta)$. If the E-space were discrete and were to consist of the states $E_1, E_2, \ldots, E_i, \ldots$, then it would be possible to define a function $\pi(n_1, n_2, \ldots, n_i, \ldots m_1, m_2 \ldots, m_i, \ldots, t)$ representing the probability that there are n_i electrons and m_i photons in the energy state $E_i(i = 1, 2, \ldots)$. Such a description of the "state" of the system is adequate enough to render the basic stochastic process defined by the random variables Markovian. Since the E-space is a continuum, no such π function can be defined unless we resort to functional techniques. Even in such a case, there is difficulty due to the situation that the initial condition on the functional cannot be uniquely defined consistently with the experimentally measurable quantities at any thickness.

3.1. METHOD OF REGENERATION POINT

The difficulties explained above can be overcome if we confine our attention to a shower excited by a single primary of specified energy. Janossy [12] observed this fact in 1950, and many of the pioneering results on the fluctuation of the number of particles are direct consequences of the basic equations for π-functions proposed by him in the historic Edinburgh Conference on Cosmic Rays.* The method of obtaining the basic equations, which is now known as the regeneration point method, has been subsequently identified (see Bartlett and Kendall [13]) with that proposed by Bellman and Harris [14] and earlier by Palm (see Harris [6], p. 131). However, Janossy's very idea of defining a suitable π-function depending on energy E is laudable

* The author is thankful to Professor Ramakrishnan for informing him of the deliberations of the Conference.

since it has led to the explicit estimation of the moments of the number distribution at any t thereby resolving difficulties considered almost pathological at that time.

Following Janossy let us define $\Phi_i(E, n; t, E_0)$ as the probability that at thickness t there are n electrons each with an energy greater than E, E_0 being the energy of the primary at $t = 0$; $i = 1$ and $i = 2$ refer respectively to the shower excited by an electron and a photon. We next observe that somewhere between 0 and t (this is called a regeneration point) the incident electron may radiate a photon (the incident photon may be annihilated forming a pair of electrons) and the electron and the photon (the two electrons) together become independent primaries each giving rise to a shower of particles. Estimating the probabilities for these "events," we obtain

(3.1)
$$\Phi_i(E, n; t, E_0) = \sum_{n_1+n_2=n} \int_0^t dt' \int dE'\, e^{-\alpha_i(E_0)t'}\, R_i\left(\frac{E'}{E_0}\right) \Phi_1(E, n_1; t - t', E')$$

$$\Phi_{3-1}(E, n_2; t - t', E_0 - E') + e^{-\alpha_i(E_0)t}\, \delta_{n,2-i}$$

subject to the initial conditions

(3.2) $$\Phi_i(n, E; 0; E_0) = \delta_{n,2-i}\ (i = 1, 2).$$

Differentiating (3.1) with respect to t, we obtain

(3.3)
$$\frac{\partial \Phi_i(n, E; t, E_0)}{\partial t} = - \int_E^{E_0} \{\Phi_i(n, E; t, E_0)$$

$$- \sum_{n_1+n_2=n} \Phi_i(n_1, E; t, E') \Phi_{3-i}(n_2, E; t; E_0 - E')\} R_i(E'|E_0)\, dE'.$$

If we define the generating function of Φ_i by

(3.4) $$G_i(u, E; t, E_0) = \Sigma u^n \Phi_i(n, E; t, E_0),$$

equation (3.3) can be written in the form

(3.5)
$$\frac{\partial G_i(u, E; t; E_0)}{\partial t} = - \int_0^{E_0} \{G_i(u, E; t; E_0)$$

$$- G_{3-i}(u, E; t; E_0 - E') G_1(u, E; t; E')\} R_i(E'|E_0)\, dE'$$

with the initial condition

(3.6) $$G_i(u, E; 0; E_0) = u^{2-i}.$$

We wish to observe $\alpha_1(E_0) = \infty$ if we use the fully screened Bethe-Heitler cross-section given by (2.1) and (2.2) and in view of this difficulty mathematicians are not happy over the equation (3.1). However, it is not very difficult to derive (3.5) without recourse to equation (3.1) by dealing with the random variable n and in fact Harris [15] has taken such a viewpoint and considers equation (3.5) as the basic set of equations.

3.2. MOMENTS OF THE NUMBER DISTRIBUTION

Equation (3.5) is nonlinear in G and hence may not be very convenient to deal with in general.* However, if we confine ourselves to the determination of the moments of the number distribution, the equations turn out to be linear in the highest order of the moment that is sought. Janossy and Messel have made use of this simplicity and obtained numerical estimates of the first two moments of the number distribution. Before we proceed to indicate the method of solution, we notice the homogeneous nature of the function R_i in its variables. This enables us to use the notation

$$(3.7) \qquad R_i(E'|E_0)\, dE' = R_i(\mathscr{E}')\, d\mathscr{E}', \qquad (\mathscr{E}' = E'/E_0)$$

and observe that G_i is a function of E/E_0 only so that we have

$$(3.8) \qquad G_i(u, E; t; E_0) = G_i(u, \mathscr{E}; t), \qquad (\mathscr{E} = E/E_0).$$

Next we observe that the equations for the factorial moments of the number distribution can be obtained by differentiating both sides of (3.5) with respect to u at $u = 1$. Thus the m-th factorial moments given by

$$(3.9) \qquad k_i(m, \mathscr{E}, t) = \left.\frac{\partial^m\, G_i(u, \mathscr{E}; t)}{\partial u^m}\right|_{u=1}$$

satisfy the equations

$$(3.10) \quad \frac{\partial \chi_i(m, \mathscr{E}, t)}{\partial t} = -\int_0^1 \left[k_i(m, \mathscr{E}, t) - k_1\left(m, \frac{\mathscr{E}}{\mathscr{E}'}, t\right) \right] R_i(\mathscr{E}')\, d\mathscr{E}'$$

$$+ \int_0^1 k_{3-i}\left(m, \frac{\mathscr{E}}{1-\mathscr{E}'}, t\right) R_i(\mathscr{E}')\, d\mathscr{E}'$$

$$+ \sum_{l=1}^{m-1} \int_0^1 k_1\left(l, \frac{\mathscr{E}}{\mathscr{E}'}, t\right) k_{3-i}\left(m-l, \frac{\mathscr{E}}{1-\mathscr{E}'}, t\right) R_i(\mathscr{E}')\, d\mathscr{E}'$$

with the initial conditions

$$(3.11) \qquad k_i(m, \mathscr{E}, 0) = \delta_{i1}\, \delta_{m1}.$$

* A simplified model of the problem in which $\alpha_1(E_0) = \alpha_2(E_0)$ (each being finite) has been dealt with by Ramakrishnan [16] who has obtained an explicit solution for $G_i(u, E; t; E_0)$.

Janossy and Messel [10] have dealt with (3.10) by taking a Mellin transform of k_i with respect to $\mathscr{E}$ and solving the resulting set of equations. If we put $m = 1$, the equations reduce to those of Landau and Rumer [17] depicting the mean behavior of the cascades. The great virtue of the set of equations (3.10) lies in its capability of explicit solution of higher moments successively starting from the first moment. Though this point has been sufficiently emphasized by Janossy [12] and explicitly demonstrated by Janossy and Messel [10] at least for the second moment, there appears to be a general (mistaken!) impression that (3.5) is intractable and recourse must be had to other methods like the correlation functions in energy variables (see, for example, Bharucha-Reid [11], in particular p. 263).

Janossy and Messel [18] and Messel [19] have prepared extensive tables of the first two moments of the number distribution based on the equations (3.10) and this will be discussed in the next chapter.

4. Product Density Approach

The product density method treated in detail in Chapter 4 has proved to be useful in the study of the stochastic features of the cascade phenomena. For instance, *if the initial conditions are given in the form of a spectrum, the problem cannot be formulated in terms of either the Janossy G equation or the characteristic functional.* In fact, from the data on the cosmic ray experiments, we can extract only mean values in the form of energy spectrum. In such a situation the product density approach is extremely useful since the initial spectrum can be expressed in terms of the initial conditions on the product densities of various orders. Such a kind of formulation is very valuable particularly if we study the development of cosmic ray showers as a whole starting from a primary nucleon. We shall discuss this aspect of the problem when we deal with nucleon cascades and extensive air showers in Chapter 7.

Following the notation of Chapter 3, let us define $f_n(E_1, E_2, \ldots, E_n; t)$ and $g_n(E_1, E_2, \ldots, E_n; t)$ as the product densities of degree n of electrons and photons respectively. We can use the symbol $fg_{i,n-i}$ to denote a mixed product density of degree n with i electrons and $n - i$ photons. Since a knowledge of the product densities of degree $\leqslant n$ will lead us to the n-th moment of the number distribution, the motivation will be to obtain a convenient set of equations satisfied by the product densities.

4.1. SUB-MARKOVIAN NATURE

The product densities of degree one are nothing but the differential mean numbers in the energy variable introduced by Bhabha and Heitler [3] and

Landau and Rumer [17]. If we denote the random variable representing the number of electrons and photons in the range $(E, E + dE)$ by $dN_i(E, t)$ ($i = 1$ and $i = 2$ will stand for electrons and photons respectively), then we have

(4.1)
$$f_1(E, t) \, dE = E\{dN_1(E, t)\}$$

$$g_1(E, t) \, dE = E\{dN_2(E, t)\}.$$

Let us assume that there is an electron of energy between E and $E + dE$ at thickness $t + \Delta$ (and this happens with probability $f_1(E, t + \Delta) \, dE$). There are three possibilities:*

(i) this electron has existed at t with a different energy E' and in traversing through Δ has dropped down to an energy between E and $E + dE$ after radiating a quantum (this happens with probability $R_1(E|E') \, dE\Delta$)

(ii) this electron has existed at t with an energy E and has continued to be in the same state without suffering any radiation—this happens with probability

$$\left(1 - \Delta \int_0^E R_1(E'|E) \, dE'\right)$$

(iii) the electron has been created between t and $t + \Delta$ by a photon of energy E' (this happens with the probability $2R_2(E|E') \, dE\Delta$).**

Taking into account these possibilities and the fact that $f_1(E, t) \, dE$ denotes the probability that there is an electron in the energy range $(E, E + dE)$, we obtain

(4.2)
$$\frac{\partial f_1(E, t)}{\partial t} = -f_1(E, t) \int_0^E R_1(E'|E) \, dE' + \int_E^\infty f_1(E', t)R_1(E|E') \, dE'$$

$$+ 2 \int_E^\infty g_1(E', t) \, R_2(E|E') \, dE'.$$

* We shall work under Approximation A.
** The factor 2 appears since we do not distinguish between electrons and positrons.

In a similar manner, we can show that g_1 satisfies the equation

$$(4.3) \quad \frac{\partial g_1(E, t)}{\partial t} = -g_1(E, t) \int_0^E R_2(E'|E)\, dE'$$

$$+ \int_E^\infty f_1(E', t)\, R_1(E' - E|E')\, dE'.$$

Higher order product densities can be shown to satisfy similar types of equations. For example, in the case of second order product densities, we obtain a set of simultaneous integrodifferential equations involving $f_2, g_2, fg_{1,1}, f_1$ and g_1. In the present case, there is a contribution from the first order densities since a photon of appropriate energy at t can give rise to a pair of electrons in its passage through $(t, t + \Delta)$ and likewise an electron will give rise to an electron-photon combination. These equations have been adequately dealt with in the monographs of Ramakrishnan [2] and Bharucha-Reid [11].

There is one characteristic property of the product densities displayed by equations (4.2) and (4.3) which has somehow escaped the attention of the earlier authors. Though the cascade process is not Markovian in nature (the process being regenerative if the shower were to be excited by a single primary or given number of primaries), a property very close to that is depicted by equations (4.2) and (4.3). The mean spectrum at $t + \Delta$ is completely determined by the mean spectrum at t. This property is shared by the product densities of higher orders as well. We can express this in a slightly different way. The correlations of the particles in energy space up to a given order at $t = 0$ determine the correlations for $t > 0$ up to the same order. We would like to place emphasis on the phrase "up to the given order," since for example the mean spectrum at $t = 0$ will not determine the mean square number of particles for $t > 0$. We shall call processes having the above characteristic as *sub-Markovian processes*, since the differential equations given above, though deceptively of Kolmogorov type, do not imply Markovian nature of the process as a whole.

4.2. HIGHER MOMENTS OF THE NUMBER DISTRIBUTION

As we have observed earlier, equations (4.2) and (4.3) have been the basis for the understanding of the mean behavior of showers excited by a single primary electron or photon. These equations were further studied by Bhabha and Chakrabarti [20, 21] by taking into account collision loss suffered by electrons. However, much reliance cannot be placed on the mean behavior without some good estimate of the spread about the mean. This was first noted by Scott and Uhlenbeck [8] who introduced some correlation functions very similar to the product densities. Their estimate of the fluctuation about the mean has been further refined by Bhabha and Ramakrishnan [9] who obtained the differential equations satisfied by the product densities of degree two and expressed the mean square number explicitly, as a double Mellin integral. In a subsequent paper, Bhabha [22] showed how the higher order correlations determine the higher moments of the number distribution and thereby provide a completion of the stochastic treatment of electron-photon cascades. This contribution of Bhabha taken along with that of Ramakrishnan [23] constitutes the complete mathematical solution of the electromagnetic cascades.

4.3. JANOSSY FUNCTIONS

The Janossy densities discussed in Chapters 3 and 4 have also been used in the study of cascade showers, particularly in the development of nucleon cascade theory. Their importance was first noted by Scott [24] who attempted to derive the higher moments of the distribution with the help of the Janossy densities. In that process, he involuntarily provided an alternative method of arriving at the results of Bhabha [22] pertaining to the cumulant moments of the number distribution. Messel and his collaborators have made a series of contributions to the analytical solution of the cascade equations based on the work of Janossy and Scott. In these contributions, they have made an extensive use of the transform technique and matrix methods and obtained explicit Mellin transform solution for the higher moments of the number distribution. For a detailed account, the reader is referred to the monograph of Bharucha-Reid [11].

5. New Approach to Cascade Theory

The cascade theory of showers deals with the number of particles at a given depth t, assuming the shower to be initiated by a particle at $t = 0$. Stated in the most general terms, the object of the cascade theory is to obtain the function $\pi^i(n_1, n_2, \ldots, n_m; E_1, E_2, \ldots, E_m | E_0, t)$ denoting the probability that there exist n_j particles of type j each of which has an energy greater than $E_j (j = 1, 2, \ldots, m)$ at t if the cascade is initiated by a particle of type i and energy E_0. What we have dealt with so far relates to the attempts to obtain integral or differential equations for π^i and the moments of π^i therefrom. This procedure is natural since electromagnetic cascades were first observed in cloud chambers where the experimental techniques made energy measurements feasible at a particular depth. However, with the advent of nuclear emulsions, it is more convenient to estimate the energies of particles at the points of their production rather than of all the particles at any particular thickness (see, for example, reference [25]). To deal with such a situation, Ramakrishnan and Srinivasan [26] have introduced the function denoting the probability that n_j particles of type j each with energy greater than E_j at the point of its production (called the primitive energy) are produced between 0 and t in a shower initiated by a primary of energy E_0 and type j. In the next sub-sections, we shall deal with the problems arising out of this new approach to cascade theory.

5.1. PRODUCTION PRODUCT DENSITIES

For preciseness, we shall confine ourselves to the development of the electromagnetic cascade. Let $dM_1(E, t)$ and $dM_2(E, t)$ denote the random variables denoting respectively the number of electrons and photons that are produced between t and $t + dt$ with primitive energies lying in the interval $(E, E + dE)$. If $\pi_1(n)$ is the probability n electrons are produced between t and $t + dt$ each with primitive energy between E and $E + dE$, then we assume as in the case of instant product densities that $\pi_1(n)$ satisfies the conditions

(5.1)*
$$\pi_1(1) = E\{dM_1(E, t)\} = F_1(E, t)\, dE\, dt$$
$$\pi_1(n) = o(dE\, dt) \qquad\qquad n > 1$$
$$\pi_1(0) = 1 - F_1(E, t)\, dE\, dt + o(dE\, dt).$$

We also assume that $\pi_2(n)$ likewise satisfies the same conditions as $\pi_1(n)$. From this point onward, it is clear that the product density technique for the product space Ω of E and t can be carried in toto. For example, the r-th moment of the number of electrons produced between 0 and t, the primitive energy of each of the electrons being greater than E_c, is given by

(5.2)
$$\mathscr{E}\{[n_1(E_c, t)]^r\} = \sum_s C_s^r \int_{E_c}^{\infty} \int_{E_c}^{\infty} \cdots \int_{E_c}^{\infty} \int_0^t \int_0^t \cdots \int_0^t$$
$$F_s(E_1, E_2, \ldots, E_s; t_1, t_2, \ldots, t_s)$$
$$dE_1\, dt_1\, dE_2\, dt_2 \cdots dE_s\, dt_s$$

where C_s^r are the coefficients introduced in Section 2 of Chapter 3 and

* Throughout, we shall use capital symbols for the corresponding product density functions in the new approach.

$F_s(E_1, E_2, \ldots, E_s; t_1, t_2, \ldots, t_s)$ is the production product density of degree s defined by

(5.3) $\quad F_s(E_1, E_2, \ldots, E_s; t_1, t_2 \ldots, t_s) \, dE_1 \, dt_1 \, dE_2 \, dt_2 \cdots dE_s \, dt_s$
$$= E\{dM_1(E_1, t_1) \, dM_1(E_2, t_2) \cdots dM_1(E_s, t_s)\}$$

where the ranges $d\Omega_1, d\Omega_2, \ldots, d\Omega_s$ in the product space of E and t do not overlap.

5.2. MEAN BEHAVIOR OF CASCADES

We shall first deal with the mean behavior of electron-photon cascades. Since the condition has a dominant role in the equations satisfied by the product densities, we shall use the superscript 1 or 2 according to whether an electron or a photon of energy E_0 is incident at $t = 0$ and write $F_1(E, t)$ and $G_1(E, t)$ as $F_1^i(E|E_0, t)$ and $G_1^i(E|E_0, t)$. Next we observe that an electron can be produced between t and $t + dt$ only by a photon, which may either be the primary or have been produced somewhere between 0 and t. We can argue in a similar way for the production of a photon between t and $t + dt$ to obtain

(5.4) $\quad F_1^i(E|E_0; t) = 2 \int\limits_0^t \int\limits_E^{E_0} G_1^i(E'|E_0; \tau) \, e^{-D(t-\tau)} \, R_2(E|E') \, dE' \, d\tau$

$$+ 2\delta_{i2} R_2(E|E_0)$$

(5.5) $\quad G_1^i(E|E_0; t)$

$$= \int\limits_E^{E_0} dE' \int\limits_{E'}^{E_0} \int\limits_0^t F_1^i(E'|E_0; \tau)\pi(E''|E'; t - \tau) \, R_1(E'' - E|E'') \, dE'' \, d\tau$$

$$+ \delta_{i1} R(E_0 - E|E_0)$$

where

(5.6) $$D = \int\limits_0^E R_2(E'|E_0) \, dE'$$

and $\pi(E''|E'; t) \, dE''$ denotes the probability that an electron of energy E' in traversing a thickness t drops to an energy between E'' and $E'' + dE''$. Defining the Mellin transform of $F_1^i(E|E_0; t)$ and $G_1^i(E|E_0; t)$ as

(5.7) $$P_1^i(s|E_0; t) = \int\limits_0^\infty F_1^i(E|E_0; t) \, E^{s-1} \, dE$$

$$(5.8) \qquad Q_1^i(s|E_0; t) = \int_0^\infty G_1^i(E|E_0; t)\, E^{s-1}\, dE$$

we obtain

$$(5.9) \qquad P_1^i(s|E_0; t) = B(s) \int_0^t Q_1^i(s|E_0; \tau)\, e^{-D(t-\tau)}\, d\tau + \delta_{i2}\, B(s)$$

$$(5.10) \quad Q_1^i(s|E_0; t) = C(s) \int_0^t P_1^i(s|E_0; \tau)\, p(s|E_0; t-\tau)\, E_0^{-s+1}\, d\tau + \delta_{i1}\, C(s)$$

where $p(s|E_0; t)$, the Mellin transform of $\pi_1(E|E_0; t)$, is given by (see reference [27])

$$(5.11) \qquad p(s|E_0; t) = E_0^{s-1}\, e^{-A(s)t},$$

and $A(s)$, $B(s)$, and $C(s)$ are defined by

$$(5.12) \quad A(s) = \left(\frac{4}{3} + \alpha\right)\left\{\frac{d}{ds}\log\Gamma(s) + \gamma - 1 + \frac{1}{s}\right\} + \frac{1}{2} - \frac{1}{s(s+1)}$$

$$(5.13) \quad B(s) = 2\left\{\frac{1}{s} - \left(\frac{4}{3} + \alpha\right)\frac{1}{(s+1)(s+2)}\right\}$$

$$(5.14) \quad C(s) = \frac{1}{s+1} + \left(\frac{4}{3} + \alpha\right)\frac{1}{s(s-1)}$$

$$(5.15) \qquad D = \frac{7}{9} - \frac{\alpha}{6}$$

where α is a constant depending on the material.

Equations (5.9) and (5.10) can be solved either by taking a Laplace transform with respect to t or by converting the integral equations into a pair of differential equations and solving the resulting system directly. We shall take up the discussion of the solution in the next chapter.

In principle, the cascade theory can be built up in terms of the production product densities, without recourse to the densities introduced in the previous section. (Of course, as is to be expected, the complexity of the equations increases with the order of the product densities.) This has been done by K. S. S. Iyer [28] who has written down a chain of integral equations connecting the various product densities of degree two and obtained the Mellin transform solution of the resulting equations. There is yet another approach which consists in expressing the production product densities in terms of the instant product densities and in fact the first numerical results [26, 29] based on the new approach were obtained only in this manner.

5.3. JANOSSY G-EQUATIONS AND MOMENTS

Let $\pi_i(n, E, E_0; t)$ be the probability that n electrons are produced between 0 and t by a primary of energy E_0. As usual, we shall assume that the probability per unit thickness of matter that (i) an electron of energy E radiates a quantum and drops to an energy between E' and $E' + dE'$ is $R_1(E'|E)\,dE'$ and (ii) a photon of energy E annihilates into an electron-positron pair one of which has an energy between E' and $E' + dE'$ is $R_2(E'|E)\,dE'$. We shall take only these two processes into account and neglect collision loss. In view of the homogeneous nature of the cross-sections (2.1) and (2.2), $\pi_i(n, E, E_0; t)$ is a function only of E/E_0 and denoting it by $\pi_i(n, \mathscr{E}, t)$ $(\mathscr{E} = E/E_0)$ we obtain

$$(5.16) \quad \frac{\partial \pi_1(n, \mathscr{E}, t)}{\partial t} = -\pi_1(n, \mathscr{E}, t) \int_0^t R_1(\mathscr{E}')\,d\mathscr{E}'$$

$$+ \sum_{m=0}^{\infty} \int_0^1 R_1(\mathscr{E}')\,\pi_1\!\left(m, \frac{\mathscr{E}}{\mathscr{E}'}, t\right) \pi_2\!\left(n - m, \frac{\mathscr{E}}{1 - \mathscr{E}'}, t\right) d\mathscr{E}'$$

$$(5.17) \quad \frac{\partial \pi_2(n, \mathscr{E}, t)}{\partial t} = -\pi_2(n, \mathscr{E}, t) \int_0^1 R_2(\mathscr{E}')\,d\mathscr{E}'$$

$$+ \int_0^{\mathscr{E}} R_2(\mathscr{E}')\,\pi_2\!\left(n - 1, \frac{\mathscr{E}}{1 - \mathscr{E}'}, t\right) d\mathscr{E}'$$

$$+ \sum_{m=0}^{\infty} \int_{\mathscr{E}}^{1-\mathscr{E}} R_2(\mathscr{E}')\,\pi_1\!\left(m, \frac{\mathscr{E}}{\mathscr{E}'}, t\right) \pi_2\!\left(n - m - 2, \frac{\mathscr{E}}{1 - \mathscr{E}'}, t\right) d\mathscr{E}'$$

$$+ \int_{1-\mathscr{E}}^1 R_2(\mathscr{E}')\,\pi_1\!\left(n - 1, \frac{\mathscr{E}}{\mathscr{E}'}, t\right) d\mathscr{E}'$$

with the initial conditions

$$(5.18) \qquad\qquad \pi_1(n, \mathscr{E}, 0) = \pi_2(n, \mathscr{E}, 0) = \delta_{n0}.$$

We observe that (5.16) holds good for the entire range $0 \leqslant \mathscr{E} \leqslant 1$, while (5.17) is valid only for the range $0 \leqslant \mathscr{E} \leqslant \frac{1}{2}$. For $\mathscr{E} > \frac{1}{2}$, π_2 satisfies the equation

$$(5.17') \quad \frac{\partial \pi_2(n, \mathscr{E}, t)}{\partial t} = -\pi_2(n, \mathscr{E}, t) \int_0^1 R_2(\mathscr{E}')\,d\mathscr{E}' + \delta_{n0} \int_{1-\mathscr{E}}^{\mathscr{E}} R_2(\mathscr{E}')\,d\mathscr{E}'$$

$$+ \int_0^{1-\mathscr{E}} R_2(\mathscr{E}')\,\pi_1\left(n-1, \frac{\mathscr{E}}{1-\mathscr{E}'}, t\right) d\mathscr{E}'$$

$$+ \int_{\mathscr{E}}^1 R_2(\mathscr{E}')\,\pi_1\left(n-1, \frac{\mathscr{E}}{\mathscr{E}'}, t\right) d\mathscr{E}'.$$

Comparing equations (5.16) and (5.17) with the corresponding ones in reference [12] we find (5.16) is identical with the equation of Janossy. The difference is brought out by (5.17) where linear terms are integrated over partial ranges. This is due to the situation that at the regeneration point one or two electrons with energy above $\mathscr{E}$ are produced. If $\mathscr{E}' = E'/E_0$ falls in the interval $(\mathscr{E}, 1 - \mathscr{E})$ then both the electrons produced have an energy above $\mathscr{E}$ and the third term on the right-hand side of (5.17) corresponds to this situation. If, however, $\mathscr{E}'$ falls outside the interval $(\mathscr{E}, 1 - \mathscr{E})$, only one of the electrons has an energy above $\mathscr{E}$ and this is taken care of by the second and the last terms in (5.17). Defining $G_i(u, \mathscr{E}, t)$ as

$$(5.19) \quad G_i(u, \mathscr{E}, t) = \sum_{n=0}^{\infty} u^n \pi_i(n, \mathscr{E}, t)$$

we obtain

$$(5.20) \quad \frac{\partial G_i(u, \mathscr{E}, t)}{\partial t} = -G_i(u, \mathscr{E}, t) \int_0^1 R_i(\mathscr{E}')\,d\mathscr{E}'$$

$$+ u^{i-1} \int_0^{\mathscr{E}} G_{3-i}\left(u, \frac{\mathscr{E}}{1-\mathscr{E}'}, t\right) R_i(\mathscr{E}')\,d\mathscr{E}'$$

$$+ u^{2i-2} \int_{\mathscr{E}}^{1-\mathscr{E}} G_1\left(u, \frac{\mathscr{E}}{\mathscr{E}'}, t\right) G_{3-i}\left(u, \frac{1-\mathscr{E}'}{\mathscr{E}}, t\right) R_i(\mathscr{E}')\,d\mathscr{E}'$$

$$+ u^{i-1} \int_{1-\mathscr{E}}^1 G_1\left(u, \frac{\mathscr{E}}{\mathscr{E}'}, t\right) R_i(\mathscr{E}')\,d\mathscr{E}'.$$

Equation (5.20) is again valid for the entire range of $\mathscr{E}$ only for $i = 1$. When $i = 2$, (5.20) covers only the region $0 < \mathscr{E} \leqslant \frac{1}{2}$. For $\mathscr{E} > \frac{1}{2}$, $G_2(u, \mathscr{E}, t)$ satisfies the equation

$$(5.20') \quad \frac{\partial G_2(u, \mathscr{E}, t)}{\partial t} = -G_2(u, \mathscr{E}, t) \int_0^1 R_2(\mathscr{E}') \, d\mathscr{E}'$$

$$+ \int_{1-\mathscr{E}}^{\mathscr{E}} R_2(\mathscr{E}') \, d\mathscr{E}' + u \int_0^{1-\mathscr{E}} R_2(\mathscr{E}') G_1 \left(u, \frac{\mathscr{E}}{1-\mathscr{E}'}, t \right) d\mathscr{E}'$$

$$+ u \int_{\mathscr{E}}^1 R_2(\mathscr{E}') \, G_1 \left(u, \frac{\mathscr{E}}{\mathscr{E}'}, t \right) d\mathscr{E}'.$$

The m-th moment of the number of electrons produced between 0 and t is given by

$$(5.21) \quad E\{[n_i(\mathscr{E}, t)]^m\} = \left(u \frac{\partial}{\partial u} \right)^m G_i(u, \mathscr{E}, t)|_{u=1}.$$

Differentiating both sides of (5.19) with respect to u at $u = 1$, we obtain

$$(5.22)^* \quad \frac{\partial n_i(\mathscr{E}, t)}{\partial t} = -n_i(\mathscr{E}, t) \int_0^1 R_i(\mathscr{E}') \, d\mathscr{E}'$$

$$+ (i - 1) \left[\int_0^{1-\mathscr{E}} R_i(\mathscr{E}') \, d\mathscr{E}' + \int_0^{\mathscr{E}} R_i(\mathscr{E}') \, d\mathscr{E}' \right]$$

$$+ \int_{\mathscr{E}}^1 R_i(\mathscr{E}') \, n_1 \left(\frac{\mathscr{E}}{\mathscr{E}'}, t \right) d\mathscr{E}' + \int_0^{1-\mathscr{E}} R_i(\mathscr{E}') \, n_{3-i} \left(\frac{\mathscr{E}}{1-\mathscr{E}'}, t \right) d\mathscr{E}'.$$

Writing

$$(5.23) \quad \frac{\partial^2 G_i(u, \mathscr{E}, t)}{\partial u^2} \bigg|_{u=1} = N_i(\mathscr{E}, t)$$

we obtain by double differentiation of (5.19) at $u = 1$

$$(5.24) \quad \frac{\partial N_1(\mathscr{E}, t)}{\partial t} = -N_1(\mathscr{E}, t) \int_0^1 R_1(\mathscr{E}') \, d\mathscr{E}' + \int_{\mathscr{E}}^1 R_1(\mathscr{E}') N_1 \left(\frac{\mathscr{E}}{\mathscr{E}'}, t \right) d\mathscr{E}'$$

$$+ \int_0^{1-\mathscr{E}} R_1(\mathscr{E}') N_2 \left(\frac{\mathscr{E}}{1-\mathscr{E}'}, t \right) d\mathscr{E}' + L_1(\mathscr{E}, t),$$

* For convenience of notation we shall drop the expectation symbol E.

$$(5.25) \quad \frac{\partial N_2(\mathscr{E}, t)}{\partial t} = -N_2(\mathscr{E}, t) \int_0^1 R_2(\mathscr{E}')\, d\mathscr{E}' + \int_{\mathscr{E}}^1 R_2(\mathscr{E}')\, N_1\left(\frac{\mathscr{E}}{\mathscr{E}'}, t\right) d\mathscr{E}'$$

$$+ \int_0^{1-\mathscr{E}} R_2(\mathscr{E}')\, N_1\left(\frac{\mathscr{E}}{1-\mathscr{E}'}, t\right) d\mathscr{E}' + L_2(\mathscr{E}, t)$$

where $L_1(\mathscr{E}, t)$ and $L_2(\mathscr{E}, t)$ are given by

$$(5.26) \quad L_1(\mathscr{E}, t) = 2 \int_0^1 n_1\left(\frac{\mathscr{E}}{\mathscr{E}'}, t\right) n_2\left(\frac{\mathscr{E}}{1-\mathscr{E}'}, t\right) R_1(\mathscr{E}')\, d\mathscr{E}'$$

$$(5.27) \quad L_2(\mathscr{E}, t) = 2 \int_0^1 n_1\left(\frac{\mathscr{E}}{\mathscr{E}'}, t\right) n_1\left(\frac{\mathscr{E}}{1-\mathscr{E}'}, t\right) R_2(\mathscr{E}')\, d\mathscr{E}'$$

$$+ 2 \int_{\mathscr{E}}^1 \left[n_1\left(\frac{\mathscr{E}}{\mathscr{E}'}, t\right) + n_1\left(\frac{\mathscr{E}}{1-\mathscr{E}'}, t\right) \right] R_2(\mathscr{E}')\, d\mathscr{E}'$$

$$+ 2 \int_0^{1-\mathscr{E}} \left[n_1\left(\frac{\mathscr{E}}{\mathscr{E}'}, t\right) + n_1\left(\frac{\mathscr{E}}{1-\mathscr{E}'}, t\right) \right] R_2(\mathscr{E}')\, d\mathscr{E}'$$

$$+ 2 \int_{\mathscr{E}}^{1-\mathscr{E}} R_2(\mathscr{E}')\, d\mathscr{E}' \left[1 - H\left(\mathscr{E} - \frac{1}{2}\right) \right]$$

where $H(\lambda)$ is the Heaviside unit function. Equation (5.21) can be easily solved by Mellin transform technique. Defining

$$(5.28) \quad p_i(s, t) = \int_0^1 \mathscr{E}^{s-1} n_i(\mathscr{E}, t)\, d\mathscr{E}$$

we obtain

$$(5.29) \quad \frac{\partial p_1(s, t)}{\partial t} = -A(s + 1)\, p_1(s, t) + C(s + 1)\, p_2(s, t)$$

$$(5.30) \quad \frac{\partial p_2(s, t)}{\partial t} = -D p_2(s, t) + B(s + 1)\, p_1(s, t) + \frac{B(s + 1)}{s}$$

where A, B, C, D are given by (5.12) through (5.15) (see, for example, reference [5]).

Equations (5.28) and (5.29) can be solved by the use of matrix calculus (see, for example, Bellman [30]). The explicit solution is given by

$$(5.31) \quad p_1(s, t) = \frac{B(s + 1) \, C(s + 1)}{s\{\mu(s + 1) - \lambda(s + 1)\}} \left[\frac{1 - e^{-\lambda(s+1)t}}{\lambda(s + 1)} - \frac{1 - e^{-\mu(s+1)t}}{\mu(s + 1)} \right]$$

$$(5.32) \quad p_2(s, t) = \frac{B(s + 1)}{s\{\mu(s + 1) - \lambda(s + 1)\}} \left[\frac{\mu(s + 1) - D}{\lambda(s + 1)} \{1 - e^{\lambda(s+1)t}\} \right.$$

$$\left. + \frac{D - \lambda(s + 1)}{\mu(s + 1)} \{1 - e^{-\mu(s+1)t}\} \right]$$

where $\lambda(s)$ and $\mu(s)$ are eigen values of the matrix

$$\begin{bmatrix} -A(s) & C(s) \\ B(s) & -D \end{bmatrix}.$$

The mean number of electrons produced between 0 and t is given by

$$(5.33) \qquad n_1(\mathscr{E}, t) = \frac{1}{2\pi i} \int_{\sigma-i\infty}^{\sigma+i\infty} \mathscr{E}^{-s} \, p_1(s, t) \, ds$$

$$(5.34) \qquad n_2(\mathscr{E}, t) = \frac{1}{2\pi i} \int_{\sigma-i\infty}^{\sigma+i\infty} \mathscr{E}^{-s} \, p_2(s, t) \, ds$$

in agreement with the results obtained using product density techniques. The numerical evaluation $n_1(\mathscr{E}, t)$ for small values of t will be discussed in the next chapter.

To obtain the mean square number we must deal with (5.24) and (5.25). Defining $P_i(s, t)$ as the Mellin transform of $N_i(\mathscr{E}, t)$ with respect to $\mathscr{E}$ we have

$$(5.35) \quad \frac{\partial P_1(s, t)}{\partial t} = -A(s + 1) \, P_1(s, t) + C(s + 1) \, P_2(s, t) + L_1(s, t),$$

$$(5.36)^* \quad \frac{\partial P_2(s, t)}{\partial t} = -D \, P_2(s, t) + B(s + 1) \, P_1(s, t) + L_2(s, t).$$

We observe that (5.35) and (5.36) are similar to (5.29) and (5.30), the difference

---

* We have used the same symbol $L_i(s,t)$ for the Mellin transform of $L_i(\mathscr{E},t)$.

being only in the inhomogeneous terms. The complete solution for P_1 and P_2 can be written in the matrix notation as

$$(5.37) \quad \begin{bmatrix} P_1(s, t) \\ P_2(s, t) \end{bmatrix} = M(s + 1)\Lambda(s + 1, t) \int_0^t \Lambda^{-1}(s + 1, t')M^{-1}(s + 1) \begin{bmatrix} L_1(s, t') \\ L_2(s, t') \end{bmatrix} dt'$$

where M and Λ are defined by

$$(5.38) \quad M(s) = \begin{bmatrix} -C(s) & -C(s) \\ -A(s) + \lambda(s) & -A(s) + \mu(s) \end{bmatrix}$$

$$(5.39) \quad \Lambda(s, t) = \begin{bmatrix} e^{-\lambda(s)t} & 0 \\ 0 & e^{-\mu(s)t} \end{bmatrix}.$$

Finally the mean square number of electrons with energy above $\mathscr{E}$ can be be expressed as

$$(5.40) \quad E\{[n_i(\mathscr{E}, t)]^2\} = E\{n_i\mathscr{E}(, t)\} + \frac{1}{2\pi i} \int_{\sigma - i\infty}^{\sigma - i\infty} P_i(s, t)\mathscr{E}^{-s}\, ds.$$

We shall take up the explicit estimation of the moments in the next chapter.

6. Invariant Imbedding Technique

The invariant imbedding technique is a mathematical tool specially designed to study the stochastic processes in a variety of natural phenomena ranging from cosmic rays to carcinogenesis. The study of the origin and evolution of this tool, by itself, constitutes an interesting piece of investigation. Invariant imbedding is an extension of certain "invariance principles" implicitly used by Sir George Stokes [31] and Lord Rayleigh [32] in the study of reflection and transmission of light by piles of plates. Later, these principles were enunciated in great detail and integrated into the general theory of radiative transfer by Ambarzumian [33] and Chandrasekhar [34]. On the other hand, the notion of imbedding has been used by Kolmogorov (see Feller [35]) in his formulation of backward and forward differential equations for Markov processes. However, the particular version of the technique, as it is known to and practiced by the applied mathematicians during the past few years, can be traced to a fundamental paper by Bellman and Harris [14] dealing with an age dependent branching phenomenon. In this contribution, Bellman and Harris have demonstrated the possibility of dealing with a non-Markovian process of a certain "duration" by imbedding it into a class of processes of arbitrary duration, the result of such an imbedding yielding a functional integral equation. This has been further developed by Bellman, Kalaba, and Wing [36–38] in their comprehensive study of neutron transport theory. Preisendorfer [39, 40] has also further developed the theory by drawing certain generalized imbedding relations. The utility of this technique in neutron fission processes has been demonstrated by Bellman *et al.* [41, 42]. In this section, we shall deal with other applications of the imbedding technique.

As we have mentioned in Section 3, Janossy [12] has taken the lead among the cosmic ray theorists by using the idea of a regeneration point to obtain an equation for the probability frequency function for the number of particles at a thickness t each having an energy greater than E. Even though Janossy's method was identified with the method of Bellman and Harris and the backward differential equation of Kolmogorov (see reference [13]) the usefulness of the approach in the product density formalism could not be brought out in a vivid manner until the formulation of the invariant imbedding technique in the present form.

6.1. IMBEDDING EQUATIONS FOR PRODUCT DENSITIES*

The product density techniques that have been developed so far adequately explain the method of arriving at the fluctuation in the size of cascade showers. Besides affording a method of solving an almost pathological problem the product density techniques proved to be very effective in describing the moments of the number distribution of particles distributed in various ranges of energy. Though the equations were written down for the number distribution of electrons and photons in an electromagnetic cascade, the methods are equally applicable to any branching process involving very many species of particles. The general method consists in defining various types of product densities and obtaining the differential equations satisfied by them. Since the product density functions, by their very definition, are independent of the distribution in numbers and energy of all the types and numbers of particles other than those contemplated in their definition, it is not difficult to obtain the differential equations satisfied by these by expressing the density function at $t + \Delta$ in terms of those at t, through integrodifferential equations [9]. To be very specific let us consider the various n-th order product

* We here follow the treatment of Ramakrishnan, Srinivasan, and Vasudevan [43].

densities encountered in cascades. The system of differential equations for these n-th order product densities arrived at by the above method (known as the last collision method in literature) can be written as

(6.1) $$\frac{\partial}{\partial t}[\mathbf{F}_n] = [A][F_n] + \mathbf{\psi}$$

where the vector $\mathbf{F}_n$ denotes the different product densities of degree n, of electrons and photons, and $\mathbf{\psi}$ is a vector involving product densities of degree $(n-1)$, A being a $2^n \times 2^n$ matrix. Since the n-th order system involves product densities of degree $(n-1)$* the equations have to be solved recursively. However, with increasing order of n, the equations become unwieldy since the order of the matrix that is encountered increases by corresponding powers of 2, thus rendering the actual computation of moments greater than two, if not impossible, at least formidably difficult. An examination of the above method will reveal that an attempt to write down the equation satisfied by the product density of degree n, involving correlation of n electrons, forces us to introduce product densities of mixed type involving the correlation of m electrons and $n - m$ photons. However, for the moment formula for the electrons we need only the product densities involving the correlation of electron alone. The complicated structure of the $2^n \times 2^n$ matrix arises from the introduction of product density of other types into the equations. However, it is possible to avoid the mixed densities if we employ the invariant imbedding techniques, by studying the possible outcome of events in the initial interval 0 to Δ of the interval 0 to t, rather than trying to express the state of the system at $t + \Delta$ in terms of that at t. The possible events occurring between 0 and Δ can only be a bremsstrahlung for an electron-initiated shower and a pair production for a photon-initiated shower or the electron and photon being constrained to move without any change in their initial states.** We observe that in this case it is sufficient to use a vector equation involving two components only (as constrasted with 2^n components in the previous case), the different components arising from different initial primaries. This is a decided advantage over the earlier method especially in view of the fact that the 2×2 matrix is the same whatever be the order of the product density involved.

We now proceed to the derivation of the imbedding relations satisfied by the product densities. Let $f_1^i(E, t|E_0)$ be the product density of degree one of electrons that are emitted in a shower, excited by a primary of i-th type having an energy E_0 ($i = 1$ and 2 signify an electron and a photon, respectively). As the

* This is on the assumption that bremsstrahlung and pair production is the cause of the development of the shower.
** For simplicity, we do not contemplate any ionization loss.

parent electron enters the medium, either nothing happens in the initial interval and the same electron impinges with same energy E_0 on the rest of the medium to produce the shower later, or it may radiate a photon of energy $E_0 - E'$ and become an electron of energy E', both of which can now act as two different parents of the shower in the rest of the medium. Translating these possibilities into mathematical terms, we write down the imbedding equations as

$$(6.2) \quad f_1^1(E, t + \Delta|E_0) = (1 - \Delta\!\int R_1(E'|E_0)\, dE')\, f_1^1(E, t|E_0)$$

$$+\Delta\!\int R_1(E'|E_0)[f_1^1(E, t|E') + f_1^2(E, t|E_0 - E')]\, dE'$$

and

$$(6.3) \quad f_1^2(E, t + \Delta|E_0) = (1 - \Delta\!\int R_2(E'|E_0)\, dE')\, f_1^2(E, t|E_0)$$

$$+2\Delta\!\int R_2(E'|E_0)\, f_1^1(E, t|E')\, dE'.$$

Hence going to the limit $\Delta \to 0$, the integrodifferential equations for first order product densities can be obtained as

$$(6.4) \quad \frac{\partial f_1^1(E, t|E_0)}{\partial t} = -f_1^1(E, t|E_0) \int R_1(E'|E_0)\, dE'$$

$$+ \int R_1(E'|E_0)[f_1^1(E, t|E') + f_1^2(E, t|E_0 - E')]\, dE',$$

$$(6.5) \quad \frac{\partial f_1^2(E, t|E_0)}{\partial t} = 2\!\int R_2(E'|E_0)\, f_1^1(E, t|E')\, dE'$$

$$-f_1^2(E, t|E_0) \int R_2(E'|E_0)\, dE'.$$

Since the cross-sections R_1 and R_2 are homogeneous in E and E', it is not difficult to see that we can express the functions $f_1^1(E, t|E_0)$ and $f_1^2(E, t|E_0)$ as functions of $E/E_0 = \mathscr{E}$ and re-express the above equations as

$$(6.6) \quad \frac{\partial f_1^1(\mathscr{E}, t)}{\partial t} = -f_1^1(\mathscr{E}, t) \int_0^1 R_1(\mathscr{E}')\, d\mathscr{E}'$$

$$+ \int_\mathscr{E}^1 R_1(\mathscr{E}') \left[\frac{1}{\mathscr{E}'} f_1^1\left(\frac{\mathscr{E}}{\mathscr{E}'}, t\right) + \frac{1}{1-\mathscr{E}'} f_1^2\left(\frac{\mathscr{E}}{1-\mathscr{E}'}, t\right) \right] d\mathscr{E}'$$

$$(6.7) \quad \frac{\partial f_1^2(\mathscr{E}, t)}{\partial t} = -\int_0^1 d\mathscr{E}'\, R_2(\mathscr{E}')\, f_1^2(\mathscr{E}, t) + 2\!\int_\mathscr{E}^1 R_2(\mathscr{E}')\, f_1^1\left(\frac{\mathscr{E}}{\mathscr{E}'}, t\right) \frac{d\mathscr{E}'}{\mathscr{E}'}.$$

Taking the Mellin's transforms of the equations with respect to $\mathscr{E}$, we obtain the matrix equation

$$(6.8) \qquad \frac{\partial}{\partial t} = \begin{bmatrix} f_1^1(s, t) \\ f_1^2(s, t) \end{bmatrix} = \begin{bmatrix} -A(s) & C(s) \\ B(s) & -D \end{bmatrix} \begin{bmatrix} f_1^1(s, t) \\ f_1^2(s, t) \end{bmatrix}$$

with initial conditions

$$(6.9) \qquad f_1^1(s, 0) = 1, \qquad f_1^2(s, 0) = 0$$

where

$$(6.10) \quad f_1^1(s, t) = \int_0^1 f_1^1(\mathscr{E}, t)\, \mathscr{E}^{s-1}\, d\mathscr{E}, \qquad f_1^2(s, t) = \int_0^1 f_1^2(\mathscr{E}, t)\, \mathscr{E}^{s-1}\, d\mathscr{E}$$

and $A(s)$, $B(s)$, $C(s)$, and D are defined by (5.12) through (5.15). Equation (6.8) is identical with the Landau-Rumer equations. This feature has been observed by Bartlett [44] in the case of nucleon cascades. However, the situation is different if we proceed to higher order product densities.

If $f_2^1(E_1, E_2, t|E_0)$ and $f_2^2(E_1, E_2, t|E_0)$ are the product densities of electrons in the electron-initiated and photon-initiated showers respectively, arguing as before, we obtain

$$(6.10) \quad \frac{\partial f_2^1(E_1, E_2, t|E_0)}{\partial t} = -f_2^1(E_1, E_2, t|E_0) \int R_1(E'|E_0)\, dE'$$
$$+ \int R_1(E'|E_0)[f_2^1(E_1, E_2, t|E')$$
$$+ f_2^2(E_1, E_2, t|E_0 - E')]\, dE'$$
$$+ \int R_1(E'|E_0)[f_1^1(E_1, t|E')f_1^2(E_2, t|E_0 - E')$$
$$+ f_1^2(E_1, t|E_0 - E')f_1^1(E_2, t|E')]\, dE',$$

$$(6.11) \quad \frac{\partial f_2^2(E_1, E_2, t|E_0)}{\partial t} = -f_2^2(E_1, E_2, t|E_0) \int R_2(E'|E_0)\, dE'$$
$$+ \int R_2(E'|E_0)[f_2^1(E, E_2, t|E')$$
$$+ f_2^1(E_1, E_2, t|E_0 - E')]\, dE'$$
$$+ 2\int R_2(E'|E_0) f_1^1(E_1, t|E')f_1^1(E_2, t|E_0 - E')\, dE'.$$

In view of the homogeneous nature of the cross-sections as before we can treat f_2^1 and f_2^2 as functions of $\mathscr{E}_1$ and $\mathscr{E}_2$ where $\mathscr{E}_1 = E_1/E_0$ and $\mathscr{E}_2 = E_2/E_0$.

Taking Mellin transforms with respect to both $\mathscr{E}_1$ and $\mathscr{E}_2$ we obtain the equations

$$(6.12) \quad \frac{\partial f_2^1(s_1, s_2, t)}{\partial t} = -f_2^1(s_1, s_2; t)\left[\int_0^1 R_1(\mathscr{E}')(1 - \mathscr{E}'^{s_1+s_2-2})\, d\mathscr{E}'\right]$$

$$+ \int_0^1 R_1(\mathscr{E}')(1 - \mathscr{E}')^{s_1+s_2-2}\, d\mathscr{E}'[f_2^1(s_1, s_2, t)$$

$$+ f_2^2(s_1, s_2, t)]$$

$$+ f_1^1(s_1, t)f_1^2(s_2, t)\int_0^1 R_1(\mathscr{E}')(1 - \mathscr{E}')^{s_2-1}\mathscr{E}'^{s_1-1}\, d\mathscr{E}'$$

$$+ f_1^2(s_1, t)f_1^1(s_2, t)\int_0^1 R_1(\mathscr{E}')(1 - \mathscr{E}')^{s_1-1}\mathscr{E}'^{s_2-1}\, d\mathscr{E}'$$

$$(6.13) \quad \frac{\partial f_2^2(s_1, s_2, t)}{\partial t} = -f_2^2(s_1, s_2, t)\int_0^1 R_2(\mathscr{E}')\, d\mathscr{E}'$$

$$+ f_2^1(s_1, s_2, t)\left[\int_0^1 R_2(\mathscr{E}')\,\mathscr{E}'^{s_1+s_2-2}\, d\mathscr{E}'\right.$$

$$\left. + \int_0^1 R_2(\mathscr{E}')(1 - \mathscr{E}')^{s_1+s_2-2}\, d\mathscr{E}'\right]$$

$$+ 2f_1^1(s_1, t)f_1^2(s_2, t)\int_0^1 R_2(\mathscr{E}')\,\mathscr{E}'^{s_1-1}(1 - \mathscr{E}')^{s_2-1}\, d\mathscr{E}'.$$

The above can be embodied in a matrix equation of the type

$$(6.14) \quad \frac{\partial}{\partial t}\begin{bmatrix} f_2^1(s_1, s_2, t) \\ f_2^2(s_1, s_2, t) \end{bmatrix}$$

$$= \begin{bmatrix} -A(s_1 + s_2 - 1) & C(s_1 + s_2 - 1) \\ B(s_1 + s_2 - 1) & -D \end{bmatrix}\begin{bmatrix} f_2^1(s_1, s_2, t) \\ f_2^2(s_1, s_2, t) \end{bmatrix}$$

$$+ \begin{bmatrix} f_1^1(s_1, t)f_1^1(s_2, t)\, a_2(s_1, s_2) + f_1^1(s_2, t)f_1^2(s_1, t)\, a_2(s_2, s_1) \\ f_1^1(s_1, t)f_1^1(s_2, t)\, a_1(s_1, s_2) \end{bmatrix}$$

with the initial condition

(6.15) $f_2^1(s_1, s_2, 0) = f_2^2(s_1, s_2, 0) = 0$

where

(6.16) $a_2(s_1, s_2) = \int R_1(\mathscr{E}') \, \mathscr{E}'^{s_1-1}(1 - \mathscr{E}')^{s_2-1} \, d\mathscr{E}'$

$a_1(s_1, s_2) = 2 \int R_2(\mathscr{E}') \, \mathscr{E}'^{s_1-1} (1- \mathscr{E}')^{s_2-1} \, d\mathscr{E}'.$

In the above matrix equation, the second vector on the right-hand side is completely known from the solutions of equations (6.8) solved previously. Also the matrix on the right-hand side is identical with the matrix that occurs in the first order equations. It is the decided advantage of this method that we are only solving a 2×2 matrix equation instead of the 4×4 matrix equation to which we are led if we employ the last collision method to write down the cascade equations. It is relevant to point out here that even if we write the equations for higher order densities, according to the imbedding technique employed here we will encounter vector matrix equations, where the matrix will be the same 2×2 matrix and the inhomogeneous vector term will include the transforms of lower order densities. The n-th order product densities, likewise, satisfy the equations:

(6.17) $\dfrac{\partial f_n^1(E_1, E_2, \ldots, E_n, t \mid E_0)}{\partial t}$

$$= -f_n^1(E_1, E_2, \ldots, E_n, t \mid E_0) \int R_1(E' \mid E_0) \, dE'$$

$$+ \int R_1(E' \mid E_0)[f_n^1(E_1, E_2, \ldots, E_n, t \mid E') + f_n^2(E_1, E_2, \ldots, E_n, t \mid E_0 - E')] \, dE'$$

$$+ \int R_1(E' \mid E_0)[\sum_{i=1}^{n-1} f_i^1(E_1, E_2, \ldots, E_i, t \mid E') f_{n-i}^2(E_{i+1}, \ldots, E_n, t \mid E_0 - E')$$

$$+ f_i^2(E_1, E_2, \ldots, E_i, t \mid E_i - E') f_{n-i}^1(E_{i+1}, \ldots, E_n, t \mid E')] \, dE',$$

(6.18) $\dfrac{\partial f_n^2(E_1, E_2, \ldots, E_n, t \mid E_0)}{\partial t} = -f_n^2(E_1, E_2, \ldots, E_n, t \mid E_0) \int R_2(E' \mid E_0) \, dE'$

$$+ \int R_2(E' \mid E_0)[f_n^1(E_1, E_2, \ldots, E_n, t \mid E')$$

$$+ f_n^1(E_1, E_2, \ldots, E_n, t \mid E_0 - E')] \, dE'$$

$$+ 2\int R_2(E' \mid E_0)[\sum_{i=1}^{n-1} f_i^1(E_1, E_2, \ldots, E_i, t \mid E')$$

$$f_{n-i}^1(E_{i+1}, \ldots, E_n, t \mid E_0 - E')] \, dE'.$$

Taking the n-fold Mellin transforms with respect to $\mathscr{E}_1, \mathscr{E}_2, \ldots, \mathscr{E}_n$ these equations become

$$(6.19) \quad \frac{\partial}{\partial t} \begin{bmatrix} f_n^1(s_1, s_2, \ldots, s_n, t) \\ f_n^2(s_1, s_2, \ldots, s_n, t) \end{bmatrix}$$

$$= \begin{bmatrix} -A(s_1 + s_2 + \ldots + s_n - n + 1) & C(s_1 + s_2 + \ldots + s_n - n + 1) \\ B(s_1 + s_2 + \ldots + s_n - n + 1) & -D \end{bmatrix}$$

$$\begin{bmatrix} f_n^1(s_1, s_2, \ldots, s_n, t) \\ f_n^2(s_1, s_2, \ldots, s_n, t) \end{bmatrix} + \psi$$

where

$$(6.20)$$

$$\psi =$$

$$\begin{bmatrix} \sum_{i=1}^{n-i} [a_2(s_1 + s_2 + \cdots + s_i - i + 1, s_{i+1} + s_{i+2} + \cdots + s_n - n + i + 1) \\ f_i^1(s_1, s_2, \ldots, s_i, t) f_{n-i}^2(s_{i+1}, s_{i+2}, \ldots, s_n, t) \\ + a_2(s_{i+2} + s_{i+2} + \cdots + s_n - n + i + 1, s_1 + s_2 + \cdots + s_i - i + 1) \\ f_i^2(s_1, s_2, \ldots, s_i, t) f_{n-i}^1(s_{i+1}, s_{i+2}, \ldots, s_n, t)] \\ \sum_{i=1}^{n-1} a_1(s_1 + s_2 + \cdots + s_i - i + 1, s_{i+1} + s_{i+2} + \cdots + s_n - n + i + 1) \\ f_i^1(s_1, s_2, \ldots, s_i, t) f_{n-i}^1(s_{i+1}, s_{i+2}, \ldots, s_n, t) \end{bmatrix}$$

An interesting feature of the above equation is the persistence of the functions a_1 and a_2 besides A, B, C, and D with suitable changes in the arguments only. Another noteworthy feature of the above equation is the transparent manner in which the dependence of the correlation functions on the "bi-Mellin transforms" of the basic cross-section is brought out, without which the solutions of the equations will be trivial.

We wish to point out that equations of this type have been formulated for the limited case of nucleon cascade by Bartlett [44]. However, the distinction between the two methods of approach could not be noticed since the nucleon cascade can be described by a scalar equation. The striking difference of the dimensionality of the vector systems in the two cases is intimately connected with the situation that the two systems satisfy different initial conditions. This can be made a little more apparent if we attempt to obtain the equations satisfied by the product densities connected with all other correlations. For any other correlation the n-th degree involving m photons and $n - m$ electrons will also satisfy the same equations (6.17) and (6.18). Thus the 2^n such systems of two component equations are equivalent to the two identical systems of differential equations containing 2^n components, but satisfying different initial conditions according to whether the shower is initiated by an electron or by a photon primary, the equivalence arising from the equality of the number of elements of the total systems.

6.2. PRODUCTION PRODUCT DENSITIES

Let $f_1(\mathscr{E}, t)$ and $f_2(\mathscr{E}_1, \mathscr{E}_2, t_1, t_2)$ be the product densities of degree one and two of electrons produced in a shower produced by an electron of energy E_0. As before, $\mathscr{E}_1, \mathscr{E}_2$ denote the ratios of the secondary energies to that of the primary. We assume only pair production by photons and bremsstrahlung by electrons and neglect ionization loss. The corresponding densities in a photon-initiated shower will be denoted by $g_1(\mathscr{E}, t)$ and $g_2(\mathscr{E}_1, \mathscr{E}_2, t_1, t_2)$. To obtain the differential equations satisfied by f and g's, we analyze the possible events in the first interval $(0, \Delta)$ rather than in $(t, t + \Delta)$. The primary electron (photon) may radiate a photon and drop down to a lower energy (create an electron-positron pair) in the interval $(0, \Delta)$ thus giving rise to two independent primaries. To incorporate the contribution from such an outcome of events, we use the invariant imbedding technique. Thus we obtain

$$(6.21) \quad \frac{\partial f_1(\mathscr{E}, t)}{\partial t} = -f_1(\mathscr{E}, t) \int_0^1 R_1(\mathscr{E}') \, d\mathscr{E}'$$

$$+ \int_0^1 R_1(\mathscr{E}') \left[f_1\left(\frac{\mathscr{E}}{\mathscr{E}'}, t\right) \frac{1}{\mathscr{E}'} + g_1\left(\frac{\mathscr{E}}{1 - \mathscr{E}'}, t\right) \frac{1}{1 - \mathscr{E}'} \right] d\mathscr{E}',$$

$$(6.22) \quad \frac{\partial g_1(\mathscr{E}, t)}{\partial t} = -g_1(\mathscr{E}, t) \int_0^1 R_2(\mathscr{E}') \, d\mathscr{E}' + 2 \int_0^1 R_2(\mathscr{E}') f_1\left(\frac{\mathscr{E}}{\mathscr{E}'}, t\right) \frac{d\mathscr{E}'}{\mathscr{E}'},$$

$$(6.23) \quad \left(\frac{\partial}{\partial t_1} + \frac{\partial}{\partial t_2}\right) f_2(\mathscr{E}_1, \mathscr{E}_2, t_1, t_2) = -f_2(\mathscr{E}_1, \mathscr{E}_2, t_1, t_1) \int_0^1 R_1(\mathscr{E}') \, d\mathscr{E}'$$

$$+ \int_0^1 R_1(\mathscr{E}') \left[f_2\left(\frac{\mathscr{E}_1}{\mathscr{E}'}, \frac{\mathscr{E}_2}{\mathscr{E}'}, t_1, t_2\right) \frac{1}{\mathscr{E}'^2} \right.$$

$$\left. + g_2\left(\frac{\mathscr{E}_1}{1 - \mathscr{E}'}, \frac{\mathscr{E}_2}{1 - \mathscr{E}'}, t_1, t_2\right) \frac{1}{(1 - \mathscr{E}')^2} \right] d\mathscr{E}'$$

$$+ \int_0^1 R_1(\mathscr{E}') \left[f_1\left(\frac{\mathscr{E}_1}{\mathscr{E}'}, t_1\right) g_1\left(\frac{\mathscr{E}_2}{1-\mathscr{E}'}, t_2\right) \right.$$

$$\left. + f_1\left(\frac{\mathscr{E}_2}{\mathscr{E}'}, t_2\right) g_1\left(\frac{\mathscr{E}_1}{1-\mathscr{E}'}, t_1\right) \right] \frac{d\mathscr{E}'}{\mathscr{E}'(1-\mathscr{E}')}$$

(6.24)
$$\left(\frac{\partial}{\partial t_1} + \frac{\partial}{\partial t_2}\right) g_2(\mathscr{E}_1, \mathscr{E}_2, t_1, t_2)$$

$$= - \int_0^1 R_2(\mathscr{E}') \, d\mathscr{E}' \, g_2(\mathscr{E}_1, \mathscr{E}_2, t_1, t_2)$$

$$+ 2 \int_0^1 R_2(\mathscr{E}') f_2\left(\frac{\mathscr{E}_1}{\mathscr{E}'}, \frac{\mathscr{E}_2}{\mathscr{E}'}, t_1, t_2\right) \frac{d\mathscr{E}'}{\mathscr{E}'^2}$$

$$+ 2 \int_0^1 R_2(\mathscr{E}') f_1\left(\frac{\mathscr{E}_1}{1-\mathscr{E}'}, t_1\right) \frac{d\mathscr{E}'}{\mathscr{E}'(1-\mathscr{E}')} f_1\left(\frac{\mathscr{E}_2}{\mathscr{E}'}, t_2\right).$$

Now we observe

(6.25)
$$N_1(\mathscr{E}, t) = \int_0^t \int_{\mathscr{E}}^1 f_1(\mathscr{E}', t') \, d\mathscr{E}' \, dt'$$

$$M_1(\mathscr{E}, t) = \int_0^t \int_{\mathscr{E}}^1 g_1(\mathscr{E}', t') \, d\mathscr{E}' \, dt'$$

$$N_2(\mathscr{E}, t) = \int_0^t \int_0^t \int_{\mathscr{E}}^1 \int_{\mathscr{E}}^1 f_2(\mathscr{E}'_1, \mathscr{E}'_2, t'_1, t'_2) \, d\mathscr{E}'_1 \, d\mathscr{E}'_2 \, dt'_1 \, dt'_2$$

$$M_2(\mathscr{E}, t) = \int_0^t \int_0^t \int_{\mathscr{E}}^1 \int_{\mathscr{E}}^1 g_2(\mathscr{E}'_1, \mathscr{E}'_2, t'_1, t'_2) \, d\mathscr{E}'_1 \, d\mathscr{E}'_2 \, dt'_1 \, dt'_2.$$

Integrating (6.21) and (6.22) over $\mathscr{E}$ and t in the appropriate ranges and using (6.25), we find that $N_1(\mathscr{E}, t)$ and $M_1(\mathscr{E}, t)$ satisfy equations which are identical with equation (5.22). Of course, the equivalence of the equations for the mean has never been in doubt. If we integrate (6.23) and (6.24) over

$\mathcal{E}_1$ and $\mathcal{E}_2$ and t_1 and t_2 and observe that $f_2(\mathcal{E}_1, \mathcal{E}_2, t_1, t_2)$ satisfies the boundary conditions ($t_1 < t_2$)

$$(6.26)\ f_2(\mathcal{E}_1, \mathcal{E}_2, t_1, 0) = 0, \qquad f_2(\mathcal{E}_1, \mathcal{E}_2, 0, t_2) = 0, \qquad g_2(\mathcal{E}_1, \mathcal{E}_2, t_1, 0) = 0$$

$$g_2(\mathcal{E}_1, \mathcal{E}_2, 0, t_2) = 2R_2(\mathcal{E}_1)\left[f_1\left(\frac{\mathcal{E}_2}{\mathcal{E}_1}, t_2\right) + f_1\left(\frac{\mathcal{E}_2}{1 - \mathcal{E}_1}, t_2\right)\right],$$

we obtain differential equations for $N_2(\mathcal{E}, t)$ and $M_2(\mathcal{E}, t)$ identical with (5.24) and (5.25). In fact, the equivalence of the moment equations can be established to any order. It is interesting to observe that although the two methods of approach were proposed as early as 1950, the equivalence of the results was never discussed except in Ramakrishnan and Mathews [5] where some discrepancies arising from the numerical results were attributed to the linear nature of the equations satisfied by the product densities as contrasted with the nonlinear equation satisfied by the π functions. This is due to the situation that differential equations governing the product densities were obtained by an analysis of events in the interval $(t - \Delta, t)$ while the equations governing π functions can be written down by an analysis of events in the first interval $(0, \Delta)$ only. An alternative set of equations for the product densities arising from an analysis of events in the first interval $(0, \Delta)$ has escaped notice due to the nonavailability of the powerful imbedding techniques. It is precisely the latter set of equations that make the equivalence between the two approaches clear.

7. Sequent Correlations

As we have emphasized in the introduction to this chapter, the construction of a complete theory of cascades is important from several points of view. While the basic motivation for the development of the theory is understood in elementary processes of interaction, shower theory is bringing to the fore more stochastic problems hitherto not considered. In the recent experimental findings of Zaimidoroga, Prokoshkin, and Tsupko-Sitnikov [7], the correlation of the number of electrons at different thicknesses plays a prominent role. This is quite natural since a non-Markovian stochastic process like the cascade theory can be adequately described by a knowledge of the correlations in the parameter with respect to which the process develops. The different types of correlations introduced by Ramakrishnan and Radha [45], Srinivasan and Iyer [46], and Srinivasan, Koteswara Rao, and Vasudevan [47] can be readily used to interpret many of the results of Prokoshkin *et al.* [7]. In the next few subsections, we shall outline the method of obtaining the different correlation functions.

7.1. EVOLUTIONARY SEQUENT CORRELATION

Let us consider an electron-photon shower initiated by an electron or a photon of energy E_0. We shall assume that the shower develops by pair creation by photons and bremsstrahlung by electrons. Defining $F_1^i(E_1, t_1; E_2, t|E_0)$ as the sequent correlation density of degree one of electrons in a shower initiated by a primary of i-th type ($i = 1, i = 2$, denote an electron and a photon primary respectively) we can obtain differential equations by considering the various possible outcomes of events that may happen in the infinitesimal interval $(0, \Delta)$ of t-axis. The primary electron (photon) may radiate a photon and drop down to a lower energy (create an electron-positron pair) in the interval $(0, \Delta)$, thus giving rise to two independent primaries. To incorporate the contribution from such an outcome of events, we use the invariant imbedding method. To be precise, we imbed the process corresponding to a thickness t into a class of processes obtained by letting t take any positive value.

Thus we obtain

(7.1) $F_1^i(E_1, t_1; E_2, t|E_0)$

$$= (1 - \Delta \int R_i(E'|E_0)\, dE')\, F_1^i(E_1, t_1 - \Delta; E_2, t - \Delta|E_0)$$
$$+ \int R_i(E'|E_0)\{F_1^1(E_1, t_1 - \Delta; E_2, t - \Delta|E')$$
$$+ F_1^{3-i}(E_1, t_1 - \Delta; E_2, t - \Delta|E_0 - E')\}\, dE'.$$

By making $\Delta \to 0$ we derive the differential equation

(7.2) $\left(\dfrac{\partial}{\partial t_1} + \dfrac{\partial}{\partial t}\right) F_1^i(E_1, t_1; E_2, t|E_0)$

$$= -\int R_i(E'|E_0)\{(E_1, t_1; E_2, t_2|E_0) - F_1^1(E_1, t_1; E_2, t|E')\}\, dE'$$
$$+ \int R_i(E'|E_0) F_1^{3-i}(E_1, t_1; E_2, t \mid E_0 - E')\, dE'.$$

To solve (7.2), we note the asymptotic forms of $R_i(E'|E_0)$ are given by (2.1) and (2.2). In view of the cross-section $R_i(E'|E_0)$ being homogeneous in E and E', we notice that

(7.3) $F_1^i(E_1, t_1; E_2, t|E_0)\, dE_1\, dE_2 = F_1^i(\mathscr{E}_1, t_1; \mathscr{E}_2, t)\, d\mathscr{E}_1\, d\mathscr{E}_2$

$$(\mathscr{E}_i = E_i/E_0).$$

Defining the Mellin transform of $F_1^i(\mathscr{E}_1, t_1; \mathscr{E}_2, t)$ as

(7.4) $\mathscr{P}_1^i(s_1, t_1; s_2, t) = \displaystyle\int_0^1 \int_0^1 \mathscr{E}_1^{s_1-1} \mathscr{E}_2^{s_2-1} F_1^i(\mathscr{E}_1, t_1; \mathscr{E}_2, t)\, d\mathscr{E}_1\, d\mathscr{E}_2$

we can write (7.2) in the matrix form as

$$(7.5) \qquad \left(\frac{\partial}{\partial t_1} + \frac{\partial}{\partial t}\right)\vec{\mathscr{P}}_1(s_1, t_1; s_2, t) = \mathbf{L}\vec{\mathscr{P}}_1(s_1, t_1; s_2, t)$$

where $\vec{\mathscr{P}}_1$ is a vector whose components are $\mathscr{P}_1^i(s_1, t_1; s_2, t)$ and $\mathbf{L}$ is the 2×2 matrix given by

$$(7.6) \qquad \mathbf{L} = \begin{bmatrix} -A(s_1 + s_2 - 1) & C(s_1 + s_2 - 1) \\ B(s_1 + s_2 - 1) & -D \end{bmatrix}$$

where A, B, C, D are defined as in the previous section. The initial conditions to be imposed on (7.2) are given by

$$(7.7) \qquad F_1^1(E_1, 0; E_2, t|E_0) = 0$$

$$(7.8) \qquad F_1^2(E_1, 0; E_2, t|E_0) = 2R_2(E_1|E_0)\pi(E_2|E_1, t)$$

where $\pi(E_2|E_1, t)\, dE_2$ denotes the probability that an electron of energy E_1 at $t = 0$ drops to an energy between E_2 and $E_2 + dE_2$ after traversing a thickness t. Thus the initial conditions satisfied by the systems of equations (7.5) are given by

$$(7.9a) \qquad \mathscr{P}_1^1(s_1, 0; s_2, t) = 0$$

$$(7.9b) \qquad \mathscr{P}_1^2(s_1, 0; s_2, t) = e^{-A(s_1)t} B(s_1 + s_2 - 1)$$

where we have made use of the well-known expression [27] for the Mellin transform of $\pi(E_2|E_1; t)$:

$$(7.10) \qquad \begin{aligned} \pi(s|E_1; t) &= \int E_2^{s-1} \pi(E_2|E_1; t)\, dE_2 \\ &= E_1^{s-1} e^{-A(s)t}. \end{aligned}$$

Equation (7.5) can be solved either by taking a Laplace transformation with respect to t and t_1 or by diagonalizing the matrix $\mathbf{L}$ and solving the resulting partial differential equations. As the method is straightforward, we refrain from giving the intermediate steps. The final solution is given by

$$(7.11a) \quad \mathscr{P}_1^1(s_1, t_1; s_2, t) = \frac{C(s_1 + s_2 - 1)\, B(s_1 + s_2 - 1)}{\mu(s_1 + s_2 - 1) - \lambda(s_1 + s_2 - 1)}\{e^{-\lambda(s_1+s_2-1)t_1}$$
$$-e^{-\mu(s_1+s_2-1)t}\}\, e^{-A(s_2)(t-t_1)},$$

$$(7.11b) \quad \mathscr{P}_1^2(s_1, t_1; s_2, t) = \frac{B(s_1 + s_2 - 1)}{\mu(s_1 + s_2 - 1) - \lambda(s_1 + s_2 - 1)}$$
$$\times \{[\mu(s_1 + s_2 - 1) - D]e^{-\lambda(s_1+s_2-1)t}$$
$$-[D - \lambda(s_1 + s_2 - 1)]e^{-\mu(s_1+s_2-1)t}\}e^{-A(s_2)(t-t_1)}.$$

The mean number of electrons that are produced with primitive energies greater than $\mathscr{E}_1$ times the energy of the primary between 0 and t and remain above a certain fraction of the primary energy at $t' > t$ can be calculated by inverting the equations (7.11) and integrating with respect to E_1 and t_1 over the appropriate range. The occurrence of functions with arguments $s_1 + s_2$ -1 and s_2 will simplify the evaluation of the inversion integrals to a great extent. Thus, we can conveniently study the mean numbers for small values of t'. A study of these mean numbers will give us a deeper insight into the problem since the manner in which the electrons drop down in energy as t increases is made clear.

The densities $F_1^i(E_1, t_1; t \mid E_0)$ can be arrived at by different arguments. The method is very simple and consists in expressing F_1^i in terms of the conventional product densities of degree one of electrons and photons that exist at t. Since the electron is produced between t_1 and $t_1 + dt_1$, it is necessary that there should be a photon at t_1 and it should create a pair between t_1 and $t_1 + dt_1$, one of the electrons having an energy in the prescribed range. Thus, if $g_1^i(E \mid E_0; t)$ is the product density of degree one of photons that exist at t due to a shower excited by a primary of energy E_0 and type i, then

$$(7.12) \quad F_1^i(E_1, t_1; E_2, t_2 \mid E_0) = 2 \int g_1^i(E \mid E_0; t_1) R_2(E_1 \mid E) \pi(E_2 \mid E_1; t - t_1) dE.$$

The equivalence of (7.11) and (7.12) can be verified by taking a Mellin transform of both sides of (7.12) and using well-known solutions [9] of cascade theory for $g_1^i(E \mid E_0; t)$. However, the method indicated in the earlier part of the section does not make use of the conventional product densities and illustrates the method of building up cascade theory purely in terms of the sequent correlation densities. The advantage of this method is apparent whenever we have to deal with the second and higher order densities.

The sequent product densities of Ramakrishnan and Radha prove to be useful in the estimation of correlations at different t's. The imbedding technique is applicable in the present case and we obtain equations identical with (6.23) and (6.24). The initial conditions in this problem necessarily have to be different from (6.26). Since the method of solution is identical with that of production product densities, we do not propose to discuss this any further. The interested reader is referred to the original paper of Ramakrishnan, Vasudevan, and Srinivasan [43].

7.2. THE METHOD OF CHARACTERISTIC FUNCTIONAL

Bartlett and Kendall [13] have demonstrated the utility of the characteristic functional in point processes with specific reference to applications in physics and biology. We shall show that a single appropriate characteristic functional can generate the various sequent correlations described above.

Let $dN_j(E, t)$ denote the random variable representing the number of particles each with energy in the interval $(E, E + dE)$ at thickness t ($j = 1$, $j = 2$ stand for an electron and photon respectively). We have defined the characteristic functional for a cascade generated by a primary of energy E_0 and type m as (see Section 4 of Chapter 4),

$$(7.13) \quad \Gamma_m(\theta_1, \theta_2, T, E_0) = E\{\exp i \int_0^{E_0} \int_0^T [dN_1(E, t)\, \theta_1(E, t) \\ + dN_2(E, t)\theta_2(E, t)]\, dt\}.$$

Next, we imbed the process corresponding to $[0, T]$ into a class of processes corresponding to $[\Delta, T]$ and let Δ tend to zero. Using the same arguments that lead to (7.1), we obtain

$$(7.14) \quad \frac{\partial \Gamma_m(\theta_1, \theta_2, T, E_0)}{\partial T} = \int_0^{E_0} [\Gamma_1(\theta_1, \theta_2, T, E')\, \Gamma_{3-m}(\theta_1, \theta_2, T, E_0 - E') \\ - \Gamma_m(\theta_1, \theta_2, T, E_0)]\, R_m(E'|E_0)\, dE'$$

with the boundary conditions

$$(7.15) \quad \Gamma_m(\theta_1, \theta_2, 0, E_0) = e^{i\theta_m(E_0,0)}.$$

By taking appropriate functional derivatives, we obtain the imbedding equations satisfied by the sequent product densities. These have been derived independently without resort to the characteristic functional by Ramakrishnan, Vasudevan, and Srinivasan [43]. If $\Phi_m(\theta_1, \theta_2, T, E_0)$ is the characteristic functional of electrons and photons produced between 0 and T, then as observed in Section 4 of Chapter 4, Φ_m and Γ_m are related by

$$(7.16) \quad \Phi_m(\chi_1, \chi_2, T, E_0) = \Gamma_m(\theta_1, \theta_2, T, E_0)$$

provided θ_j is chosen to satisfy the condition

$$(7.17) \quad \theta_j(E, t) = \int_0^E \chi_{3-j}(E', t)\, R_j(E'|E)\, dE'.$$

It is interesting to observe that while Φ_m satisfies the same imbedding equation as Γ_m, the initial condition to be imposed on Φ_m is to be replaced by

$$(7.18) \quad \Phi_m(\chi_1, \chi_2, 0, E_0) = \int_0^{E_0} R_m(E'|E_0)\, e^{i[\chi_{3-m}(E_0-E',0)+(m-1)\chi_{3-m}(E',0)]} dE'.$$

REFERENCES

1. L. Janossy, *Cosmic Rays*, Oxford University Press, London (1950).
2. A. Ramakrishnan, *Elementary Particles and Cosmic Rays*, Pergamon Press, London (1962).
3. H. J. Bhabha and W. Heitler, *Proc. Roy. Soc.* (London), **A 159** (1937), 432.
4. J. F. Carlson and J. R. Oppenheimer, *Phys. Rev.*, **51** (1937), 220.
5. A. Ramakrishnan and P. M. Mathews, *Prog. Theor. Phys.*, **11** (1954), 95.
6. T. E. Harris, *Branching Phenomena*, Springer-Verlag, Berlin (1963).
7. O. A. Zaimidoroga, Yu. D. Prokoshkin, and V. M. Tsupko-Sitnikov, *Zhurnal Eksper. Teoret. Fiz.*, **52** (1967), 79. Translated in *Soviet Physics—JETP*, **25** (1967), 51.
8. W. T. Scott and G. E. Uhlenbeck, *Phys. Rev.*, **62** (1942), 497.
9. H. J. Bhabha and A. Ramakrishnan, *Proc. Ind. Acad. Sci.*, **A 32** (1950), 141.
10. L. Janossy and H. Messel, *Proc. Phys. Soc.*, **63** (1950), 1101.
11. A. T. Bharucha-Reid, *Elements of the Theory of Markov Processes and Their Applications*, McGraw-Hill, New York (1960).
12. L. Janossy, *Proc. Phys. Soc.*, **63** (1950), 241.
13. M. S. Bartlett and D. G. Kendall, *Proc. Camb. Phil. Soc.*, **47** (1951), 65.
14. R. E. Bellman and T. E. Harris, *Proc. Nat. Acad. Sci. U.S.A.*, **34** (1948), 601.
15. T. E. Harris, *Proc. Nat. Acad. Sci. U.S.A.*, **43** (1957), 509.
16. A. Ramakrishnan, *J. Roy. Statist. Soc.*, **B 13** (1951), 131.
17. L. D. Landau and G. Rumer, *Proc. Roy. Soc.* (London), **166** (1938), 213.
18. Leonie Janossy and H. Messel, *Proc. Roy. Irish Acad. Sci.*, **A 54** (1951), 217.
19. H. Messel, *Proc. Phys. Soc.*, **64** (1950), 807.
20. H. J. Bhabha and S. K. Chakrabarti, *Proc. Roy. Soc.* (London), **A 181** (1943), 267.
21. H. J. Bhabha and S. K. Chakrabarti, *Phys. Rev.*, **74** (1948), 1352.
22. H. J. Bhabha, *Proc. Ind. Acad. Sci.*, **A 32** (1950), 154.
23. A. Ramakrishnan, *Proc. Camb. Phil. Soc.*, **46** (1950), 595.
24. W. T. Scott, *Phys. Rev.*, **82** (1951), 893.
25. H. Fay, *Nuovo Cimento*, **5** (1957), 293.
26. A. Ramakrishnan and S. K. Srinivasan, *Proc. Ind. Acad. Sci.*, **A 44** (1956), 263.
27. L. D. Landau, *J. Phys. (U.S.S.R.)*, **8** (1944), 201.
28. K. S. S. Iyer, *Stochastic Point Processes and Their Applications to Physical Problems*, Ph.D. Thesis, Indian Institute of Technology, Madras, 1964.
29. S. K. Srinivasan and N. R. Ranganathan, *Proc. Ind. Acad. Sci.*, **A 45** (1957), 69, 268.
30. R. E. Bellman, *Introduction to Matrix Analysis*, McGraw-Hill Co., New York, 1960.
31. George Stokes, *Mathematical and Physical Papers of Sir George Stokes*, Vol. IV, Cambridge, 1904, p. 145.
32. Rayleigh, *Scientific Papers of Lord Rayleigh*, Vol. VI, Cambridge, 1920, p. 492.
33. V. A. Ambarzumian, *J. Phys. Acad. Sci. of U.S.S.R.*, **8** (1944), 65.
34. S. Chandrasekhar, *Radiative Transfer*, Oxford (1950), p. 89.
35. W. Feller, *An Introduction to Probability Theory and Its Applications*, John Wiley, New York, 1957, Chapter 17.
36. R. E. Bellman, R. Kalaba, and G. M. Wing, *Proc. Nat. Acad. Sci. (U.S.A.)*, **43** (1957), 517.
37. R. E. Bellman, R. Kalaba, and G. M. Wing, *J. Math. and Mech.*, **7** (1958), 149.
38. R. E. Bellman, R. Kalaba, and G. M. Wing, *J. Math. Phys.*, **1** (1960), 280.
39. R. W. Preisendorfer, *Proc. Nat. Acad. Sci. U.S.A.*, **44** (1958), 320.
40. R. W. Preisendorfer, *Proc. Nat. Acad. Sci. U.S.A.*, **47** (1961), 591.

41. R. E. Bellman, R. Kalaba, and M. C. Prestrud, *Invariant Imbedding and Radiative Transfer in Slabs of Finite Thickness*, Modern Analytic and Computational Methods in Science and Mathematics, American Elsevier Publishing Co., New York, 1963.
42. R. E. Bellman, H. H. Kagiwada, R. E. Kalaba, and M. C. Prestrud, *Invariant Imbedding and Time Dependent Transport Processes*, American Elsevier Publishing Co., New York, 1964.
43. A. Ramakrishnan, S. K. Srinivasan, and R. Vasudevan, *J. Math. Analys. Applcns.*, **11** (1965), 278.
44. M. S. Bartlett, *Stochastic Processes*, University Press, Cambridge, 1955, p.82.
45. A. Ramakrishnan and T. K. Radha, *Proc. Camb. Phil. Soc.*, **57** (1961), 843.
46. S. K. Srinivasan and K. S. S. Iyer, *Nuovo Cimento*, **33** (1964), 273.
47. S. K. Srinivasan, N. V. Koteswara Rao, and R. Vasudevan, *Nuovo Cimento* (1969) (to be published).

Chapter 6

ELECTROMAGNETIC CASCADES— ANALYTIC AND COMPUTATIONAL METHODS

1. Introduction

The mathematical theory of electron-photon showers, as formulated by Bhabha and Heitler [1] and Carlson and Oppenheimer [2] explains the mean behavior of cascades. The problem of estimating the fluctuation of the shower size about the mean remained open till the emergence of new techniques due to Kendall [3], Janossy [4], Bhabha [5], and Ramakrishnan [6]. The success of these techniques has been demonstrated explicitly by Bhabha and Ramakrishnan [7] and Janossy and Messel [8]. Bhabha and Ramakrishnan used the product density techniques formulated and developed by these authors themselves while Janossy and Messel employed the π-function techniques of Janossy based on the regeneration point method. Bhabha [9] in a subsequent paper outlined the complete solution of the fluctuation problem of electron-photon showers and this marked the culmination of the pioneering efforts of Bhabha, Heitler, Carlson, and Oppenheimer since the theory can be considered to have reached a state of completion. However, in 1954, Chou and Schein [10] observed a number of anomalous showers in emulsions and Fay [11] at Göttingen reported similar showers. Though most of the showers can be fully accounted for by the theory of bremsstrahlung and pair production, the present experimental evidence cannot wholly exclude multiple processes in which there may be emission of two or more high energy quanta (see, for example, Heitler [12], p. 224) or pair production by charged particles (see Bhabha [13]). While it is necessary to derive accurate cross-sections for these higher order processes,* the cascade theory itself has to be recast in a better manner to facilitate the interpretation of the observed cosmic ray events in nuclear emulsions. A beginning in this direction has been made by Ramakrishnan and Srinivasan [16] and this has been discussed in Section 6 of the previous chapter. With the emergence of the new approach to cascade

* Some explicit estimates have been made in this direction by Murota, Ueda, and Tanaka [14], Ramakrishnan, Srinivasan, Ranganathan, and Vasudevan [15].

theory it has become possible to define a variety of sequent correlation densities, and the unification of all these densities leading to a complete and comprehensive theory of cascades has been demonstrated in the final section of the previous chapter. However, to correlate with experiments, it is necessary to have tables of the moments of the number distribution for different ranges of energy and thickness. In the present chapter, we propose to give an adequate account of the efforts made in this direction.

2.   Mean Behavior of Cascades

As observed in the Introduction, the mean number of electrons and photons at any t, each having an energy greater than E can be calculated on the basis of the solutions presented in references [1] and [2]. Bhabha and Chakrabarti [17] have tabulated the mean numbers for various ranges of energy and thickness. This was soon followed by Leonie Janossy and Messel [18] who had prepared extensive tables of the mean numbers of electrons and photons in both electron- and photon-initiated showers. They have employed the saddle point method to invert the Mellin transform of the mean numbers. We do not propose to discuss this aspect any further since the monograph of Janossy [19] contains an extensive account of the same. In the next few sub-sections we shall discuss the subsequent developments in the technique and mode of calculations.

2.1.   NUMERICAL EVALUATION OF THE INVERSION INTEGRAL

Under approximation A, the average number of particles of type i that are found at a given depth t and of energy $> E$ in a shower excited by a single particle of type j and of energy E_0 is given by (see reference [17])

$$(2.1) \qquad E\{N_j^i(E, E_0, t)\} = \frac{1}{2\pi i} \int_{\sigma-i\infty}^{\sigma+i\infty} \left(\frac{E_0}{E}\right)^{s-1} \frac{p_j^i(s, t)\, ds}{(s-1)[\mu(s) - \lambda(s)]}$$

where

$$(2.2) \qquad p_1^1(s, t) = [D - \lambda(s)]\, e^{-\lambda(s)t} + (\mu(s) - D)\, e^{-\mu(s)t}$$
$$p_1^2(s, t) = C(s)\, [e^{-\lambda(s)t} - e^{-\mu(s)t}]$$
$$p_2^1(s, t) = B(s)\, [e^{-\lambda(s)t} - e^{-\mu(s)t}]$$
$$p_2^2(s, t) = (\mu(s) - D)\, e^{-\lambda(s)t} + (D - \lambda(s))\, e^{-\mu(s)t}$$

and λ and μ are the roots of the equation

(2.3) $\qquad x^2 + x(A(s) + D) + (A(s) D - B(s) C(s)) = 0$

and A, B, C, D are defined by equations (5.12) through (5.15) of Chapter 5.

As we have mentioned in the Introduction, the tables presented in references [17] and [18] were prepared by the use of the saddle point method and were restricted to all but small thicknesses. For small thicknesses, Bhabha and Heitler [1] and Arley [20] evaluated the mean numbers directly by allowing the emission of one or two photons by the initial electron and the consequent materialization of the photons into electron pairs. While such a procedure is valid for very small thicknesses, it does not yield mean numbers of fairly good accuracy for moderately small thicknesses. The difficulty can be overcome by numerically evaluating the integral on the right-hand side of (2.1) directly. In such a case, the sequence of contributions to the integral converges slowly. Butcher, Chartres, and Messel [21] have overcome the

Table 1. Average Numbers of Electrons Produced by an Electron

$$\log_{10} E\{N_1^1 \ (E, E_0/t)\}$$

Depth t (rad. lengths)	$\varepsilon = \log_{10}(E_0/E)$									
	1	2	3	4	5	6	7	8	9	10
0.1	−0.001	0.012	0.022	0.032	0.041	0.051	0.060	0.069	0.078	0.087
0.2	0.004	0.045	0.080	0.113	0.144	0.174	0.202	0.228	0.254	0.278
0.3	0.013	0.091	0.159	0.219	0.274	0.324	0.371	0.414	0.454	0.493
0.4	0.024	0.145	0.246	0.333	0.409	0.478	0.540	0.597	0.650	0.700
0.5	0.035	0.201	0.334	0.445	0.541	0.626	0.701	0.771	0.834	0.894
0.6	0.052	0.252	0.423	0.552	0.667	0.765	0.853	0.933	1.007	1.076
0.7	0.058	0.307	0.504	0.654	0.784	0.896	0.995	1.086	1.169	1.246
0.8	0.064	0.360	0.581	0.750	0.895	1.019	1.130	1.231	1.323	1.408
0.9	0.068	0.409	0.653	0.841	0.999	1.136	1.258	1.368	1.469	1.562
1.0	0.072	0.455	0.721	0.926	1.098	1.247	1.379	1.498	1.608	1.709
1.1	0.074	0.497	0.784	1.007	1.192	1.352	1.495	1.623	1.741	1.850
1.2	0.075	0.536	0.844	1.083	1.282	1.453	1.606	1.743	1.869	1.986
1.3	0.074	0.572	0.901	1.156	1.367	1.550	1.712	1.858	1.993	2.117
1.4	0.072	0.605	0.954	1.224	1.449	1.642	1.814	1.969	2.111	2.243
1.5	0.069	0.635	1.004	1.290	1.527	1.732	1.913	2.076	2.226	2.365
1.6	0.064	0.663	1.052	1.353	1.602	1.817	2.008	2.180	2.337	2.483
1.7	0.058	0.688	1.097	1.413	1.674	1.900	2.100	2.280	2.445	2.597
1.8	0.052	0.711	1.140	1.470	1.744	1.980	2.189	2.377	2.549	2.708
1.9	0.041	0.732	1.180	1.525	1.811	2.057	2.275	2.471	2.651	2.816
2.0	0.030	0.751	1.218	1.578	1.875	2.131	2.358	2.562	2.749	2.921

Table 2. Average Numbers of Electrons Produced by a Photon

$$\log_{10} E\{N_{\frac{1}{2}} (E, E_0, t)\}$$

Depth t (rad. lengths)	$\varepsilon = \log_{10} (E_0/E)$									
	1	2	3	4	5	6	7	8	9	10
0.1	−0.890	−0.831	−0.822	−0.818	−0.814	−0.811	−0.807	−0.804	−0.800	−0.797
0.2	−0.612	−0.539	−0.519	−0.505	−0.492	−0.480	−0.467	−0.455	−0.443	−0.431
0.3	−0.458	−0.367	−0.332	−0.303	−0.277	−0.251	−0.226	−0.202	−0.179	−0.156
0.4	−0.355	−0.242	−0.188	−0.141	−0.098	−0.058	−0.020	0.017	0.052	0.085
0.5	−0.278	−0.142	−0.065	0.000	0.060	0.116	0.167	0.216	0.262	0.305
0.6	−0.208	−0.068	0.050	0.125	0.207	0.276	0.340	0.401	0.456	0.510
0.7	−0.165	0.010	0.148	0.245	0.342	0.425	0.501	0.573	0.638	0.701
0.8	−0.130	0.080	0.238	0.357	0.468	0.565	0.653	0.735	0.809	0.880
0.9	−0.101	0.142	0.323	0.461	0.588	0.697	0.796	0.887	0.971	1.050
1.0	−0.077	0.200	0.402	0.560	0.700	0.822	0.932	1.032	1.125	1.212
1.1	−0.057	0.253	0.476	0.653	0.806	0.940	1.060	1.170	1.271	1.366
1.2	−0.041	0.302	0.547	0.740	0.907	1.053	1.183	1.302	1.411	1.513
1.3	−0.028	0.347	0.613	0.824	1.004	1.160	1.300	1.428	1.546	1.655
1.4	−0.017	0.389	0.676	0.903	1.096	1.263	1.413	1.549	1.675	1.791
1.5	−0.008	0.428	0.735	0.978	1.183	1.362	1.521	1.666	1.799	1.922
1.6	−0.002	0.465	0.792	1.050	1.267	1.456	1.625	1.778	1.919	2.049
1.7	0.003	0.499	0.845	1.118	1.348	1.547	1.725	1.887	2.035	2.172
1.8	0.004	0.530	0.896	1.184	1.425	1.635	1.822	1.991	2.147	2.291
1.9	0.005	0.559	0.944	1.246	1.500	1.720	1.915	2.093	2.256	2.407
2.0	0.004	0.587	0.989	1.306	1.571	1.801	2.006	2.191	2.361	2.519

difficulty by deforming the path of integration into the parabola $y^2 = 4a$ $(\sigma − x)$. By choosing suitable values of a and σ, these authors have ensured fairly rapid convergence of the cumulative contributions to the integral. Tables 1 and 2 contain the numerical results for the average number of particles carried out on the Computer SILLIAC.

2.2. MEAN NUMBER OF ELECTRONS PRODUCED IN (0, t)

We recall from Section 5 of Chapter 5 the expressions for the mean numbers of electrons that are produced between 0 and t by a primary of i-th type:

$$(2.4) \quad n^1(y, t) = \frac{1}{2\pi i} \int_{\sigma-i\infty}^{\sigma+i\infty} \frac{B(s)\, C(s)}{\mu(s) − \lambda(s)} \left\{ \frac{1 − e^{-\lambda(s)t}}{\lambda(s)} − \frac{1 − e^{-\mu(s)t}}{\mu(s)} \right\} \frac{e^{y(s-1)}}{s − 1}\, ds$$

$$(2.5) \qquad n^2(y, t) = \frac{1}{2\pi i} \int_{\sigma-i\infty}^{\sigma+i\infty} \frac{B(s)}{\mu(s) - \lambda(s)} \left\{ \frac{\mu(s) - D}{\lambda(s)} (1 - e^{-\lambda(s)t}) \right.$$

$$\left. + \frac{D - \lambda(s)}{\mu(s)} (1 - e^{-\mu(s)t}) \right\} \frac{e^{y(s-1)}}{s - 1} \, ds.$$

For fairly large thicknesses (say $t \geq 4$), we can neglect the term containing $e^{-\mu(s)t}$ and (2.4) can be written as

$$(2.6) \qquad n^1(y, t) = \frac{1}{2\pi i} \int_{\sigma-i\infty}^{\sigma+i\infty} e^{\phi_1(s)} \, ds - \frac{1}{2\pi i} \int_{\sigma-i\infty}^{\sigma+i\infty} e^{\phi_2(s,t)} \, ds$$

where

$$(2.7) \qquad \phi_1(s) = \log \frac{B(s)\, C(s)}{\mu(s)\, \lambda(s)} - \log(s - 1) + y(s - 1)$$

$$(2.8) \quad \phi_2(s, t) = \log \frac{B(s)\, C(s)}{[\mu(s) - \lambda(s)]\lambda(s)} - \log(s - 1) + y(s - 1) - t\lambda(s).$$

We observe that as t tends to infinity the second integral tends to 0 and hence we have

$$(2.9) \qquad n^1(y, \infty) = \frac{1}{2\pi i} \int_{\sigma-i\infty}^{\sigma+i\infty} e^{\phi_1(s)} \, ds.$$

As $s \to 2$, $\lambda(s) \to 0$ and $B(s)\, C(s)/\mu(s)$ remains finite and non-zero at $s = 2$. Hence $\phi_1(s)$ and $\phi_2(s, t) \to +\infty$ as $s \to 2$. From (5.12) to (5.15) of Chapter 5 it follows that for very large real part of s,

$$(2.10) \qquad A(s) \simeq \left(\frac{4}{3} + \alpha\right)\left(\log s + \gamma - 1 + \frac{1}{2s}\right) + \frac{1}{2}$$

$$B(s) \simeq \frac{2}{s}$$

$$C(s) \simeq \frac{1}{s + 1}$$

$$\lambda(s) \simeq D - \frac{2}{(\frac{4}{3} + \alpha)\, s(s + 1)\log s}$$

$$\mu(s) \simeq A(s) + \frac{2}{(\frac{4}{3} + \alpha)\, s(s + 1)\log s}.$$

Thus

$$\log \frac{B(s)\,C(s)}{\mu(s)\,\lambda(s)} - \log(s-1)$$

and

$$\log \frac{B(s)\,C(s)}{[\mu(s) - \lambda(s)]\,\lambda(s)} - \log(s-1) - t\,\lambda(s)$$

tend to $-\infty$ logarithmically. Since $y(s-1) \to +\infty$ much more rapidly, $\phi_1(s)$ and $\phi_2(s, t)$ tend to $+\infty$, as $s \to \infty$. Hence ϕ_1 and ϕ_2 must have a minimum as s increases along the real axis from 2 to ∞. Let s_1 and s_2 be respectively the points at which

(2.11) $$\frac{d\phi_1(s)}{ds} = 0, \qquad \frac{\partial\phi_2(s, t)}{\partial s} = 0.$$

We can shift the line of integration in each of the integrals so that it passes through the saddle points. This can always be achieved since the only restriction on the line of integration is that it should be to the right of the imaginary axis and also to the right of all singularities, so that $\sigma > 2$. Having chosen the contour to pass through the saddle point, we have the well-known approximation of the saddle point method obtained by replacing each of the functions ϕ_1 and ϕ_2 by the first three terms of its Taylor expansion about the saddle point. Thus

(2.12) $$\phi_1(s) \simeq \phi_1(s_1) - \xi^2 \left(\frac{d^2\phi}{ds^2}\right) s = s_1$$

(2.13) $$\phi_2(s, t) \simeq \phi_2(s_2; t) - \eta^2 \left(\frac{\partial^2\phi_2}{\partial s^2}\right)_{s=s_2}$$

where $i\xi = s - s_1$ in (2.12) and $i\eta = s - s_2$ in (2.13). Substituting (2.12) and (2.13) in (2.6), we have

(2.14) $$n^1(y, t) = \frac{1}{2\pi} \int_{-\infty}^{+\infty} e^{\phi_1(s_1) - \frac{1}{2}\xi^2 \phi_1''(s_1)}\, d\xi - \frac{1}{2\pi} \int_{-\infty}^{+\infty} e^{\phi_2(s,t) - \frac{1}{2}\eta^2 \phi_2''(s_2,t)}\, d\eta.$$

Hence we have

(2.15) $$n^1(y, t) = \frac{e^{\phi_1(s_1)}}{\sqrt{2\pi\,\phi_1''(s_1)}} - \frac{e^{\phi_2(s,t)}}{\sqrt{2\pi\,\phi_2''(s_2, t)}}.$$

The mean numbers $n^1(y, t)$ for various values of y and t have been calculated by Srinivasan and Ranganathan [22] numerically using (2.15). In order

Table 3.   The Mean Number of Electrons That Are Produced in [0, t]
(The mean numbers of electrons that exist at t are given in parentheses.)

y	t				
	4	6	8	16	∞
8	293	528	851	1372	1406
	(94.2)	(149)	(146)		
7	124	249	355	502	508
	(48.8)	(61.3)	(49.6)		
6	64.5	111	146	183	184.5
	(23.4)	(22.9)	(15.1)		
5	30.6	41.6	57.9	66.4	66.5
	(10.2)	(7.55)	(4.01)		
4	10.9	14.7	18.9	20.51	20.53
	(3.85)	(2.11)	(1.897)		
3	6.65	7.86	8.56	8.82	8.82
	(1.21)	(0.475)	(0.160)		

Table 4
The Mean Number of Electrons That Are Produced by an Electron in [0, t]
(The mean number that exist at t is given in parentheses.)

t	y							
	4	5	6	7	8	9	10	11
0.5	0.524	0.750	0.994	1.251	1.521	1.801	2.092	2.394
	(1.139)	(1.338)	(1.553)	(1.781)	(2.019)	(2.268)	(2.527)	(2.796)
0.6	0.729	1.053	1.407	1.786	2.188	2.612	3.057	3.526
	(1.252)	(1.530)	(1.836)	(2.163)	(2.512)	(2.880)	(3.269)	(3.678)
0.7	0.961	1.400	1.888	2.418	2.988	3.598	4.248	4.940
	(1.381)	(1.751)	(2.162)	(2.611)	(3.096)	(3.616)	(4.173)	(4.767)
0.8	1.216	1.790	2.437	3.151	3.931	4.777	5.693	6.679
	(1.524)	(1.995)	(2.530)	(3.123)	(3.773)	(4.483)	(5.254)	(6.088)
0.9	1.492	2.220	3.054	3.989	5.027	6.170	7.423	8.792
	(1.677)	(2.261)	(2.936)	(3.698)	(4.548)	(5.491)	(6.530)	(7.670)
1.0	1.788	2.690	3.740	4.938	6.288	7.797	9.475	11.330
	(1.837)	(2.546)	(3.378)	(4.336)	(5.424)	(6.649)	(8.020)	(9.546)
1.1	2.103	3.199	4.497	6.003	7.726	9.682	11.887	14.360
	(2.003)	(2.846)	(3.856)	(5.039)	(6.406)	(7.970)	(9.747)	(11.752)
1.2	2.433	3.745	5.326	7.189	9.356	11.851	14.703	17.940
	(2.172)	(3.159)	(4.366)	(5.806)	(7.498)	(9.466)	(11.733)	(14.326)
1.3	2.779	4.327	6.226	8.502	11.189	14.329	17.965	22.145
	(2.343)	(3.485)	(4.908)	(6.638)	(8.705)	(11.147)	(14.002)	(17.310)

to obtain the saddle points, it was found necessary to sub-tabulate at intervals of 0.025, the basic functions $A(s)$, $B(s)$, $C(s)$, $\lambda(s)$, and $\mu(s)$ tabulated by Janossy and Messel [18] at intervals of 0.1. The mean number of electrons produced in infinite thickness and fairly large thickness are given in Table 3.

Table 5

The Mean Number of Electrons That Are Produced by a Photon in [0, t]

(The mean number that exist at t is given in parentheses.)

t	y							
	4	5	6	7	8	9	10	11
0.5	0.710	0.766	0.822	0.879	0.938	1.000	1.063	1.128
	(0.682)	(0.746)	(0.806)	(0.865)	(0.925)	(0.987)	(1.050)	(1.115)
0.6	0.863	0.951	1.044	1.141	1.243	1.349	1.459	1.573
	(0.817)	(0.917)	(1.015)	(1.114)	(1.216)	(1.323)	(1.432)	(1.546)
0.7	1.025	1.157	1.298	1.450	1.611	1.781	1.959	2.146
	(0.955)	(1.102)	(1.249)	(1.403)	(1.564)	(1.733)	(1.910)	(2.095)
0.8	1.198	1.384	1.590	1.813	2.054	2.311	2.584	2.872
	(1.097)	(1.300)	(1.511)	(1.736)	(1.975)	(2.229)	(2.499)	(2.784)
0.9	1.382	1.636	1.922	2.237	2.582	2.954	3.354	3.781
	(1.243)	(1.513)	(1.803)	(2.117)	(2.457)	(2.823)	(3.216)	(3.637)
1.0	1.579	1.914	2.298	2.729	3.206	3.727	4.293	4.906
	(1.391)	(1.741)	(2.125)	(2.550)	(3.016)	(3.526)	(4.080)	(4.680)
1.1	1.788	2.219	2.722	3.295	3.935	4.645	5.426	6.281
	(1.541)	(1.982)	(2.479)	(3.037)	(3.659)	(4.349)	(5.109)	(5.942)
1.2	2.011	2.553	3.196	3.939	4.782	5.728	6.781	7.947
	(1.694)	(2.237)	(2.864)	(3.580)	(4.392)	(5.304)	(6.322)	(7.452)
1.3	2.246	2.916	3.723	4.668	5.756	6.992	8.385	9.947
	(1.847)	(2.505)	(3.280)	(4.182)	(5.220)	(6.403)	(7.741)	(9.243)

For convenience of comparison, we have also tabulated the mean number of electrons that exist at t with an energy above E.*

For small thickness, we cannot neglect the term containing $e^{-\mu(s)t}$ in (2.4). Then the saddle point formula as adapted here cannot be applied to this term. However, the integral can be evaluated directly by deforming the straight line path from $\sigma - i\infty$ to $\sigma + i\infty$ to the parabola $y^2 = 4a(\sigma - x)$. By choosing $a = 2$, $\sigma = 1$ mean numbers of electrons have been evaluated by Srinivasan et al. [24]. Tables 4 and 5 contain the mean number of electrons produced by

* These are taken from the table of Ramakrishnan and Mathews [23].

an electron and photon primary respectively. For comparison, we have also tabulated the corresponding mean numbers that exist at t.

The experimental results obtained by Fay [11] can now be compared with the mean numbers given above. We first note that the energy referred to by Fay is the total energy of the pair and the shower is initiated by a pair of electrons obtained from photon materialization. From a physical point of

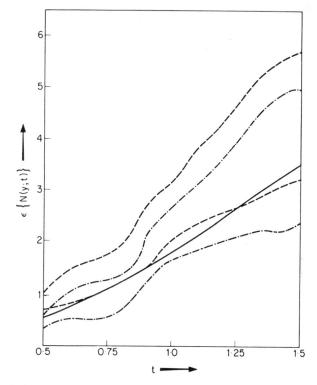

Fig. 1.   Mean number of pairs plotted against thickness measured in radiation units, $y = 4$. Broken curves denote the total number of observed pairs, while broken curves with dots denote the total number of pairs excluding tridents.

(Reproduced from Nuovo Cimento, 9, (1958) 81.)

view, it is clear that the mean number of pairs produced by a pair is exactly the same as the mean number of single electrons produced by a single electron, provided the primary electron and the pair have the same energy.*

* For a formal proof, see reference [24].

Hence the numbers presented in Table 1 can be directly compared with Fay's results. Since Fay's data are based on six showers, the statistics can be expected to be reasonably good. Further as the energies involved are fairly high, we have used the data based on scattering measurements, rather than on opening angle.

In Figures 1 and 2, we have plotted the theoretical mean numbers of pairs and the experimental limits against t for $y = 4$ and 5. For comparison, we have also indicated the extent to which the experimental curves will be

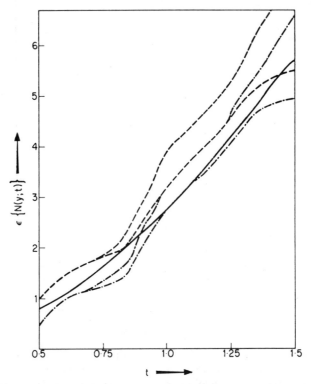

Fig. 2. Mean number of pairs plotted against thickness measured in radiation units; $y = 5$. Broken curves denote the total number of observed pairs, while broken curves with dots denote the total number of pairs excluding tridents.

(Reproduced from *Nuovo Cimento*, **9**, (1958) 81.)

depressed if the reported tridents are subtracted. It will be found that there is good agreement between the theoretical curve and the depressed curves (obtained by subtracting the tridents).

3. Higher Moments of the Number Distribution

While the mean behavior of showers is a good description of the cascade phenomena, the fluctuation about the mean should also be determined in order that we be able to decide whether an observed shower can be interpreted as a fluctuation or excluded as being highly improbable. A measure of the fluctuation is provided by the mean square deviation of the number of particles at a particular t, the energy of the particles lying in a given energy interval. The pioneering attempts in this direction by Scott and Uhlenbeck [25] yielded an estimate of the mean square number of electrons produced by an electron. In fact, Scott and Uhlenbeck anticipated the product density technique, and the origin of the moment formula developed by Bhabha [5] and Ramakrishnan [6] can be traced to this fundamental contribution. We summarize in Table 6 the main results of Scott and Uhlenbeck.

Table 6

	$\bar{n}$	$\bar{n}^2$	σ^2	$\sigma^2/\bar{n}$	$\sigma_F{}^2$	$\sigma_P{}^2$
Calculation I	1612	309.4	49 ± 16	$3.0 + 1$	243.7	16
Calculation II	17.0	0.442	153	9	272	17

$$x = \text{depth} \qquad z = \log (E_0/E) = 5.67 \qquad x = z$$

Further attempts in this direction had to be postponed till the advent of the π-function technique of Janossy [4]. As we have already discussed in Section 3 of the previous chapter, Janossy's technique provides a simple method of calculating the higher moments of the number of particles. Janossy and Messel [8] have used Mellin transform technique to study the moment equations given by (3.10) of Chapter 5.

Bhabha and Ramakrishnan [7] have used the product density to obtain explicit expressions for the mean square deviation in the form of the double Mellin inversion integrals. These double integrals have been numerically evaluated by Ramakrishnan and Mathews [23] who have obtained fairly extensive tables for the mean square and mean square deviation of the number of electrons and photons produced by an electron. Since these

numerical results have been discussed in their proper perspective in the monograph of Ramakrishnan, we wish to limit our discussion to some interesting
features.

The numerical estimate of Scott and Uhlenbeck are in agreement with the
results of Ramakrishnan and Mathews within 20 per cent. In fact, Scott and
Uhlenbeck summed up their calculations as follows. "The accuracy of the
present calculations is hard to estimate and the limits given are more or less
a guess. However, it seems sure that also in a cosmic ray case, the fluctuations
are much smaller than the Furry value and of the order of a few times the
Poisson value." This has been fully confirmed by the results of Ramakrishnan
and Mathews, which, in turn, are in good agreement with the results of
Janossy and Messel. Janossy and Messel have observed that the ratio of the
mean square deviation to the corresponding value of a Furry process is zero
and hence the showers have a tendency to a Poisson distribution near the
shower maximum. After the publication of the results by Ramakrishnan and
Mathews, Messel pointed out in a private communication that such a result
is essentially due to the result of incomplete tabulation in the original paper
of Janossy and Messel and a consequent error in their inference. Janossy and
Messel tabulated $\sigma^2/(E\{N^2\} + E\{N\})$ and wherever they obtained a value of
the order of 0.05, they inferred it was nearly 0. However, it happens that the
ratio $\sigma^2/E\{N\}$ is many times the Poisson value even at the shower maximum.

3.1. NEW APPROACH TO CASCADE THEORY—SMALL THICKNESSES

The mean square numbers of particles produced in $[0, t]$ has been explicitly
evaluated by Srinivasan and Iyer [26]. Since the experimental data on showers
usually pertain to the measurement of energies of pairs produced, it is found
convenient to deal with the number distribution of electron pairs having an
energy above E at the point of formation. The equations satisfied by the
probability frequency functions of the number of pairs can be written down
exactly in the same manner as was demonstrated in Section 5.2 of Chapter 5.
To conserve space, we give below the equations satisfied by the Mellin
transforms of the factorial moments of the number distribution of pairs
produced between 0 and t;*

$$(3.1) \qquad \frac{\partial N_1(s, t)}{\partial t} = -A(s + 1) N_1(s, t) + C(s + 1) M_1(s, t)$$

* For convenience, we have used the symbols N and M to denote respectively
the electron- and photon-initiated showers, the subscript standing for the order of
the moments.

$$(3.2) \qquad \frac{\partial M_1(s, t)}{\partial t} = -D\, M_1(s, t) + B(s + 1)\, N_1(s, t) + \frac{D}{s}$$

$$(3.3) \qquad \frac{\partial N_2(s, t)}{\partial t} = -A(s + 1)\, N_2(s, t) + C(s + 1)\, M_2(s, t) + L_1(s, t)$$

$$(3.4) \qquad \frac{\partial M_2(s, t)}{\partial t} = -D\, M_2(s, t) + B(s + 1)\, N_2(s, t) + L_2(s, t)$$

where

$$(3.5) \qquad L_1(s, t) = 2 \int_0^1 \mathscr{E}^{s-1}\, d\mathscr{E} \int_0^1 N_1\!\left(\frac{\mathscr{E}}{\mathscr{E}'}, t\right) M_1\!\left(\frac{\mathscr{E}}{1 - \mathscr{E}'}, t\right) R_1(\mathscr{E}')\, d\mathscr{E}'$$

$$(3.6) \qquad L_2(s, t) = 2 \int_0^1 \mathscr{E}^{s-1}\, d\mathscr{E} \int_0^1 N_1\!\left(\frac{\mathscr{E}}{\mathscr{E}'}, t\right) N_1\!\left(\frac{\mathscr{E}}{1 - \mathscr{E}'}, t\right) R_2(\mathscr{E}')\, d\mathscr{E}'.$$

We next solve for $N_1(s, t)$ and $M_1(s, t)$ and attempt to invert the transforms:

$$(3.7) \qquad N_1(\mathscr{E}, t) = \frac{1}{2\pi i} \int_{\sigma-i\infty}^{\sigma+i\infty} \frac{DC(s + 1)\, \mathscr{E}^{-s}}{s[\mu(s + 1) - \lambda(s + 1)]} \left[\frac{1 - e^{-\lambda(s+1)t}}{\lambda(s + 1)} \right.$$
$$\left. - \frac{1 - e^{-\mu(s+1)t}}{\mu(s + 1)} \right] ds$$

$$(3.8) \qquad M_1(\mathscr{E}, t) = \frac{1}{2\pi i} \int_{\sigma-i\infty}^{\sigma+i\infty} \frac{\mathscr{E}^{-s}}{\mu(s + 1) - \lambda(s + 1)} \left[\frac{\mu(s + 1) - D}{\lambda(s + 1)} \right.$$
$$\left. \{1 - e^{-\lambda(s+1)t}\} + \frac{D - \lambda(s + 1)}{\mu(s + 1)} \{1 - e^{-\mu(s+1)t}\} \right] ds$$

We notice that the quantities within the square bracket in the integrands multiplied by $[\mu(s + 1) - \lambda(s + 1)]^{-1}$ are analytic functions of s, the branch cuts arising from λ or μ exactly cancelling with a corresponding factor in $[\mu(s + 1) - \lambda(s + 1)]^{-1}$. Thus for small values of t, the exponential functions of λt and μt can be expanded and the resulting expansion can be algebraically expressed in terms of the functions A, B, C, D. Retaining terms of the order of t^4, we obtain after a straightforward but tedious calculation:

$$(3.9) \qquad N_1(y, t) = -0.322\, t^2 - 0.924\, t^3 + 0.016\, t^4$$
$$+ y(0.387\, t^2 - 0.110\, t^3 + 0.019\, t^4)$$
$$+ 0.088\, t^3 y^2 + e^{-y}\, [0.516\, t^2$$
$$- 0.104\, t^3 - 0.01\, t^4) + y(0.09\, t^3 - 0.033\, t^4)$$
$$- 0.014\, t^4 y^2]$$

(3.10) $M_1(y, t) = -0.774\, t - 0.299\, t^2 - 0.385\, t^3 + 0.053\, t^4$
$$+ y(0.263\, t^3 - 0.101\, t^4) + e^{-y}[(0.227\, t^3 - 0.642\, t^4)$$
$$+ y(0.35\, t^2 - 0.174\, t^4) + 0.044\, t^4\, y^2]$$

where $y = \log (E_0/E)$. The second factorial moments can be explicitly solved by matrix methods from equations (3.3) and (3.4):

(3.11) $$\begin{pmatrix} N_2(\mathscr{E}, t) \\ M_2(\mathscr{E}, t) \end{pmatrix} = \frac{1}{2\pi i} \int_{\sigma - i\infty}^{\sigma + i\infty} \begin{pmatrix} N_2(s, t) \\ M_2(s, t) \end{pmatrix} e^{y(s-1)}\, ds$$

where

(3.12) $$\begin{pmatrix} N_2(s, t) \\ M_2(s, t) \end{pmatrix} = M(s)\, \Lambda(s, t) \int_{\sigma - i\infty}^{\sigma + i\infty} \Lambda^{-1}(s, t')M(s) \begin{bmatrix} L_1(s, t') \\ L_2(s, t') \end{bmatrix} dt'$$

(3.13) $$M(s) = \begin{pmatrix} -C(s) & -C(s) \\ -A(s) + \lambda(s) & -A(s) + \mu(s) \end{pmatrix}$$

(3.14) $$\Lambda(s, t) = \begin{pmatrix} e^{-\lambda(s)t} & 0 \\ 0 & e^{-\mu(s)t} \end{pmatrix}.$$

Performing the operations implied in equation (3.11), we finally obtain

(3.15) $N_2(y, t) = t^4 (-0.248 + 0.301\, y) + \dfrac{e^{-y}\, t^4}{4} (-0.549 + 0.301\, y)$

$$+ \frac{e^{-y} t^5}{5} (-1.138 + 0.424\, y + 0.054\, y^2)$$

(3.16) $M_2(y, t) = \dfrac{2t^5}{5} (-0.193 + 0.301\, y - 0.678\, y^2)$

$$+ \frac{2t^5\, e^{-y}}{5} (0.373 - 0.752\, y + 0.15\, y^2).$$

An interesting feature of the solutions given by (3.15) and (3.16) is the absence of terms of the order lower than t^4. This, in turn, implies that the number distribution is very nearly Poissonian in the initial stages of the development of the shower. Tables 7–12 give the first two moments and the mean square deviation of the number of electron pairs produced for various values of y and t.

To compare the results with the available experimental data, we notice that a shower is detected by the formulation of the first pair and the data usually contain the energy of the first pair and the energies and positions of

Table 7.   Mean Number of Electron Pairs Produced by an Electron

y	t				
	0.1	0.2	0.3	0.4	0.5
2	0.0038	0.0129	0.0213	0.0229	0.1692
3	0.0073	0.0235	0.0518	0.0789	0.2566
4	0.0111	0.03488	0.0753	0.1129	0.3048
5	0.0149	0.04518	0.1086	0.1791	0.3820
6	0.0187	0.06268	0.1417	0.2298	0.4802
7	0.0225	0.07667	0.1649	0.2880	0.5692
8	0.0263	0.09072	0.2085	0.3467	0.6592

Table 8.   Mean Square Number of Electron Pairs Initiated by an Electron

y	t				
	0.1	0.2	0.3	0.4	0.5
2	0.0042	0.0132	0.0213	0.0269	0.1782
3	0.00731	0.0238	0.0538	0.0809	0.2606
4	0.00111	0.0351	0.0749	0.1148	0.3088
5	0.0149	0.0452	0.1095	0.1811	0.386
6	0.0187	0.0637	0.1421	0.2318	0.484
7	0.0225	0.0778	0.1654	0.2901	0.5731
8	0.0263	0.0917	0.2088	0.3487	0.663

Table 9.   Mean Square Deviation of Electron Pairs Initiated by an Electron

y	t				
	0.1	0.2	0.3	0.4	0.5
2	0.00317	0.01293	0.02086	0.02637	0.1493
3	0.00725	0.0231	0.0512	0.0745	0.1930
4	0.01008	0.03404	0.06978	0.1026	0.2182
5	0.01465	0.0432	0.0974	0.1487	0.2416
6	0.0183	0.0597	0.1225	0.1789	0.2546
7	0.0221	0.07213	0.1395	0.2060	0.2595
8	0.0257	0.08362	0.1647	0.2262	0.2621

Table 10. Mean Number of Electron Pairs Produced by a Photon

y	t				
	0.1	0.2	0.3	0.4	0.5
2	0.0743	0.1433	0.2091	0.269	0.3167
3	0.07458	0.1452	0.2161	0.2856	0.3430
4	0.0748	0.1472	0.2231	0.3022	0.3693
5	0.07514	0.1492	0.2301	0.3188	0.3956
6	0.07536	0.1509	0.2372	0.3204	0.4219
7	0.07562	0.1528	0.2443	0.3371	0.4482
8	0.0759	0.1549	0.2513	0.3536	0.4745

Table 11. Mean Square Number of Electron Pairs Initiated by a Photon

y	t				
	0.1	0.2	0.3	0.4	0.5
2	0.0743	0.1430	0.2071	0.251	0.3124
3	0.07458	0.1449	0.2141	0.2676	0.3544
4	0.0748	0.1468	0.2211	0.2843	0.3964
5	0.0751	0.1487	0.2281	0.3008	0.4384
6	0.07536	0.1506	0.2351	0.3024	0.4804
7	0.07562	0.1525	0.2421	0.319	0.5224
8	0.07588	0.1544	0.2491	0.3356	0.5644

Table 12. Mean Square Deviation of Electron Pairs Initiated by a Photon

y	t				
	0.1	0.2	0.3	0.4	0.5
2	0.0694	0.1244	0.1630	0.1791	0.2011
3	0.06968	0.02392	0.1657	0.1892	0.2388
4	0.0699	0.1263	0.1727	0.1943	0.2595
5	0.07024	0.1262	0.1752	0.1984	0.2884
6	0.07036	0.1281	0.1799	0.2001	0.314
7	0.07001	0.13	0.1845	0.2101	0.3199
8	0.07002	0.1319	0.1866	0.2131	0.3435

each one of the subsequent pairs produced in the shower excited by the first pair. Thus the functions $\pi_i(n, E, E_0, t)$ or their moments cannot be compared directly with the experimental data. However, we notice that the probability that one of the electrons of the initial pair has an energy between E_0 and $E_0 + dE_0$ is given by $R_2(E_0|E_0')\, dE_0/D$ where E_0' is the energy of the primary photon that is found to materialize into a pair. The function $\pi_i(n, E, E_0, t)$ can be viewed as the conditional probability and the corresponding moments as conditional moments. Thus we can average over all the moments by the function $R_2(E_0|E_0')/D$ and integrate over E_0 to obtain the moments of the number of pairs produced by the initial pair. Tables 13, 14, and 15 give the corresponding mean, mean square, and mean square deviation of the number of

Table 13.   Mean Number of Electron Pairs Produced by an Electron Pair

y	t				
	0.1	0.2	0.3	0.4	0.5
2	0.0177	0.0714	0.1616	0.3228	0.5845
3	0.0274	0.1111	0.2635	0.5007	0.9667
4	0.0376	0.1540	0.3654	0.6946	1.2747
5	0.0477	0.1962	0.4733	0.9145	1.6885
6	0.0585	0.2454	0.5872	1.1304	2.1481
7	0.0681	0.2869	0.7071	1.3723	2.6535
8	0.0783	0.3336	0.833	1.6302	3.2047

Table 14.   Mean Square Number of Electron Pairs Produced by an Electron Pair

y	t			
	0.1	0.2	0.3	0.4
2	0.0177	0.0721	0.1636	0.3302
3	0.0274	0.112	0.2668	0.5161
4	0.0376	0.1552	0.3702	0.7200
5	0.0477	0.1977	0.4792	
6	0.0585	0.2472	0.5944	
7	0.0681	0.2890	0.7156	
8	0.0783	0.3302	0.8478	

Table 15. Mean Square Deviation of Electron Pairs Produced by an Electron Pair

y	t			
	0.1	0.2	0.3	0.4
2	0.01746	0.0675	0.1382	0.2402
3	0.02686	0.0955	0.1935	0.3104
4	0.03624	0.1241	0.2466	0.3250
5	0.0455	0.1508	0.2677	
6	0.0547	0.1750	0.2868	
7	0.0637	0.1969	0.2939	
8	0.0726	0.2164	0.2890	

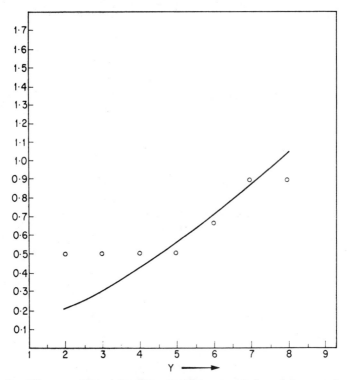

Fig. 3. Mean number of pairs of electrons produced by an electron pair corresponding to thickness $t = 0.3$. The experimental numbers are indicated by circles.

pairs. Figures 3 and 4 show the variation of the mean number of pairs with y for $t = 0.3$ and 0.4. The mean numbers calculated from the data of Fay [11] and Fenyves *et al.* [27] are indicated in the form of circles. The agreement

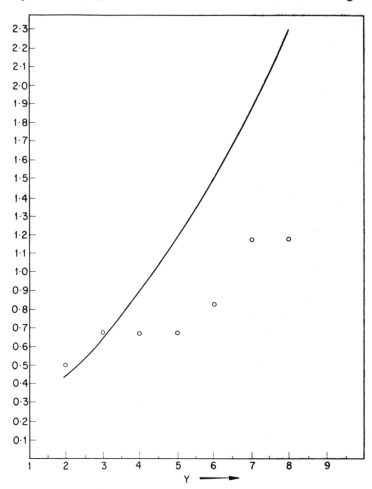

Fig. 4. Mean number of pairs of electrons produced by an electron pair corresponding to thickness $t = 0.4$. The experimental numbers are indicated by circles.

is very good for $t = 0.3$ while for $t = 0.4$, the theoretical mean numbers are higher than the corresponding experimental results. The deviation from the theoretical curve is within the limits of statistical fluctuations (see, for example, Table 15) for $t = 0.3$, while this is not so for $t = 0.4$. We also observe that we have neglected the production of tridents and the inclusion of this will still further raise the mean numbers. The disagreement is due to the fact that we have stopped with terms of the order of t^4 and this is not very satisfactory for large values of y since we encounter terms of the type $y^2 t^4$.

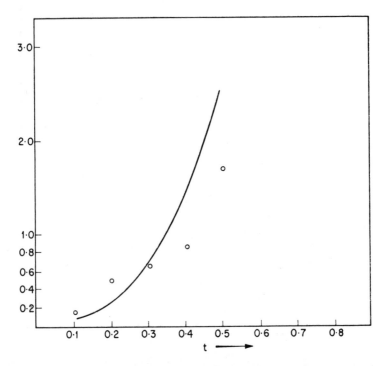

Fig. 5. Variation of the mean number of pairs with t plotted in cascade units corresponding to $y = 6$. The experimental numbers are indicated by circles

In fact, the better agreement for smaller values of y is precisely due to this situation.

Figures 5 and 6 show the variation of the mean number of pairs with t for $y = 6$ and 7. The good agreement for small values of t, especially up to $t = 0.3$ is, of course, due to the fast convergence of the expansion in power of t.

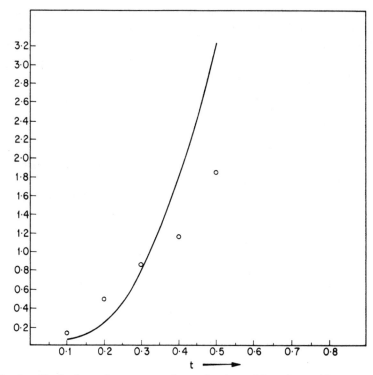

Fig. 6. Variation of mean number of pairs with t plotted in cascade units corresponding to $y = 7$. The experimental numbers are indicated by circles.

4. Distribution of the Total Number of Electrons Produced

It is well-known that the instant product densities and all the moments of the number of particles above a certain energy $E_c(E_c > 0)$ at thickness t tend to zero as t tends to infinity. This is obvious from a physical point of view. Since the Bethe-Heitler cross-sections are homogeneous the total cross-section is independent of the energy of the particle and as such the probability of finding a particle above a certain energy E_c at t tends to zero as t tends to infinity. As the production product densities are expressed as linear functionals of the instant product densities, we can assert always that for the same reason, the moments of the number of particles that are produced between 0 and t with the primitive energy of each particle being greater than E_c tends to a finite limit as t tends to infinity. The existence of these limits and their usefulness was discussed by Ramakrishnan and Srinivasan [16] on the basis of the connecting relations between the production and the instant densities. Here we shall deal with the recent results of Srinivasan *et al.* [28] who have employed the π-function technique and obtained the first few moments by inverting the Mellin transforms directly.

4.1. CASCADE EQUATIONS FOR INFINITE THICKNESS

Let $\pi_i(n, E, E_0, t)$ be the probability that n electrons above an energy E are produced between 0 and t by a primary of i-th type with energy E_0 ($i = 1, i = 2$ refer respectively to an electron and a photon). In view of the homogeneous nature of the cross-sections for pair creation by a photon and radiation of a photon by an electron $\pi_i(n, E, E_0; t)$ is only a function of E/E_0. We assume that these functions have a limit as t tends to infinity and set

$$(4.1) \qquad \lim_{t \to \infty} \pi_i(n, \mathscr{E}, t) = \pi_i(n, \mathscr{E})$$

where we denote $\pi_i(n, E, E_0, t)$ as $\pi_i(n, \mathscr{E}, t)$ $\quad \mathscr{E} = E/E_0$.

From Section 5.3 of Chapter 5, we notice that equations (5.16), (5.17), and (5.17′) become, in the limit as t tends to infinity,

$$(4.2) \quad -\pi_1(n,\mathscr{E}) \int_0^1 R_1(\mathscr{E}') \, d\mathscr{E}'$$

$$+ \sum_{m=0}^{\infty} \int_0^1 R_1(\mathscr{E}') \, \pi_1(m, \mathscr{E}/\mathscr{E}') \, \pi_2(n - m, \mathscr{E}/1 - \mathscr{E}') \, d\mathscr{E}' = 0$$

$$(4.3a) \quad \pi_2(n, \mathscr{E}) \int_0^1 R_2(\mathscr{E}') \, d\mathscr{E}' = \int_0^{\mathscr{E}} R_2(\mathscr{E}') \, \pi_1(n - 1, \mathscr{E}/1 - \mathscr{E}') \, d\mathscr{E}'$$

$$+ \sum_{m=0}^{\infty} \int_{\mathscr{E}}^{1-\mathscr{E}} R_2(\mathscr{E}') \, \pi_1(m, \mathscr{E}/\mathscr{E}') \, \pi_1(n - m - 2, \mathscr{E}/1 - \mathscr{E}') \, d\mathscr{E}'$$

$$+ \int_{1-\mathscr{E}}^{1} R_2(\mathscr{E}') \, \pi_1(n - 1, \mathscr{E}/\mathscr{E}') \, d\mathscr{E}' \qquad\qquad (0 \leqslant \mathscr{E} \leqslant 1/2)$$

$$(4.3b) \quad \pi_2(n, \mathscr{E}) \int_0^1 R_2(\mathscr{E}') \, d\mathscr{E}' = \delta_{n0} \int_{1-\mathscr{E}}^{\mathscr{E}} R_2(\mathscr{E}') \, d\mathscr{E}'$$

$$+ \int_0^{1-\mathscr{E}} R_2(\mathscr{E}') \, \pi_1(n - 1, \mathscr{E}/1 - \mathscr{E}') \, d\mathscr{E}'$$

$$+ \int_{\mathscr{E}}^1 R_2(\mathscr{E}') \, \pi_1(n - 1, \mathscr{E}/\mathscr{E}') \, d\mathscr{E}' \qquad (\tfrac{1}{2} < \mathscr{E} \leqslant 1)$$

Introducing the generating function $g_i(u, \mathscr{E})$ by

$$(4.4) \qquad\qquad g_i(u, \mathscr{E}) = \sum_n \pi_i(n, \mathscr{E}) \, u^n$$

we obtain

$$(4.5) \quad g_i(u, \mathscr{E}) \int_0^1 R_i(\mathscr{E}') \, d\mathscr{E}' = u^{i-1} \int_0^{\mathscr{E}} g_{3-i}(u, \mathscr{E}/1 - \mathscr{E}') \, R_1(\mathscr{E}') \, d\mathscr{E}'$$

$$+ u^{2i-2} \int_{\mathscr{E}}^{1-\mathscr{E}} g_1(u, \mathscr{E}/\mathscr{E}') g_{3-i}(u, \mathscr{E}/1 - \mathscr{E}') \, R_i(\mathscr{E}) \, d\mathscr{E}$$

$$+ u^{i-1} \int_{1-\mathscr{E}}^1 g_{3-i}(u, \mathscr{E}/\mathscr{E}') \, R_i(\mathscr{E}') \, d\mathscr{E}'.$$

(4.5) is valid for the entire range of $\mathscr{E}$ for $i = 1$ only. When $i = 2$, equation (4.5) covers only the range $0 \leqslant \mathscr{E} \leqslant \tfrac{1}{2}$. For $\mathscr{E} > \tfrac{1}{2}$ we have

$$(4.6) \quad g_2(u, \mathscr{E}) \int_0^1 R_2(\mathscr{E}') \, d\mathscr{E}' = \int_{1-\mathscr{E}}^{\mathscr{E}} R_2(\mathscr{E}') \, d\mathscr{E}'$$

$$+ u \int_0^{1-\mathscr{E}} R_2(\mathscr{E}') \, g_1(u, \mathscr{E}/1 - \mathscr{E}') \, d\mathscr{E}' + u \int_{\mathscr{E}}^1 R_2(\mathscr{E}') \, g_1(u, \mathscr{E}/\mathscr{E}') \, d\mathscr{E}'.$$

Equations (4.5) and (4.6) are not capable of explicit solution. However, if we restrict ourselves to the factorial moments, the problem becomes tractable. Defining the factorial moments as

$$(4.7) \qquad\qquad N_m(\mathscr{E}) = \frac{\partial^m g_1}{\partial u^m} \bigg|_{u=1}$$

$$M_m(\mathscr{E}) = \frac{\partial^m g_2}{\partial u^m} \bigg|_{u=1}$$

we find that the second factorial moments satisfy the equations

$$(4.8) \qquad -N_2(\mathscr{E}) \int_0^1 R_1(\mathscr{E}')\, d\mathscr{E}' + \int_{\mathscr{E}}^1 N_2(\mathscr{E}|\mathscr{E}')\, R_1(\mathscr{E}')\, d\mathscr{E}'$$

$$+ \int_0^{1-\mathscr{E}} M_2(\mathscr{E}|1-\mathscr{E}')\, R_1(\mathscr{E}')\, d\mathscr{E}' + 2[1-H(\mathscr{E}-\tfrac{1}{2})]$$

$$\int_{\mathscr{E}}^{1-\mathscr{E}} N_1(\mathscr{E}|\mathscr{E}')\, M_1(\mathscr{E}|1-\mathscr{E}')\, R_1(\mathscr{E}')\, d\mathscr{E}' = 0$$

$$(4.9) \quad M_2(\mathscr{E}) \int_0^1 R_2(\mathscr{E}')\, d\mathscr{E}'$$

$$= \int_0^{1-\mathscr{E}} N_2(\mathscr{E}|1-\mathscr{E}')\, R_2(\mathscr{E}')\, d\mathscr{E}' + \int_{\mathscr{E}}^1 N_2(\mathscr{E}|\mathscr{E}')\, R_2(\mathscr{E}')\, d\mathscr{E}'$$

$$+ 2\int_0^{1-\mathscr{E}} N_1(\mathscr{E}|1-\mathscr{E}')\, R_2(\mathscr{E}')\, d\mathscr{E}' + 2\int_{\mathscr{E}}^1 N_1(\mathscr{E}|\mathscr{E}')\, R_2(\mathscr{E}')\, d\mathscr{E}'$$

$$+ 2[1-H(\mathscr{E}-\tfrac{1}{2})]\int_{\mathscr{E}}^{1-\mathscr{E}} [N_1(\mathscr{E}|\mathscr{E}') + M_1(\mathscr{E}|1-\mathscr{E}') + 1$$

$$+ N_1(\mathscr{E}|\mathscr{E}')\, M_1(\mathscr{E}|1-\mathscr{E}')]\, R_2(\mathscr{E}')\, d\mathscr{E}'.$$

The first factorial moments are just the mean numbers and are given by (see, for example, reference [16])

$$(4.10) \qquad N_1(\mathscr{E}) = \frac{1}{2\pi i} \int_{\sigma-i\infty}^{\sigma+i\infty} \frac{B(s)\, C(s)}{(s-1)\, [A(s)\, D - B(s)\, C(s)]}\, e^{y(s-1)}\, ds$$

$$(4.11) \qquad M_1(\mathscr{E}) = \frac{1}{2\pi i} \int_{\sigma-i\infty}^{\sigma+i\infty} \frac{B(s)\, A(s)}{(s-1)\, [A(s)\, D - B(s)\, C(s)]}\, e^{y(s-1)}\, ds$$

where

$$(4.12) \qquad\qquad y = \log(E_0/E)$$

and A, B, C, D are defined by equations (5.12) through (5.15) of Chapter 5

To solve (4.8) and (4.9), we use the Mellin transform technique. Defining $N_2(s)$ and $M_2(s)$ by

$$(4.13) \qquad N_2(s) = \int_0^1 N_2(\mathscr{E})\, \mathscr{E}^{s-1}\, d\mathscr{E}$$

$$(4.14) \qquad M_2(s) = \int_0^1 M_2(\mathscr{E})\, \mathscr{E}^{s-1}\, d\mathscr{E}$$

we obtain

$$(4.15) \qquad A(s+1)\, N_2(s) - C(s+1)\, M_2(s) = L_1(s+1)$$

$$(4.16) \qquad -B(s+1)\, N_2(s) - D\, M_2(s) = L_2(s+1)$$

where

$$(4.17a) \quad L_1(s+1) = \int_0^{1/2} 2\mathscr{E}^{s-1}\, d\mathscr{E} \int_{\mathscr{E}}^{1-\mathscr{E}} N_1(\mathscr{E}/\mathscr{E}')\, M_1(\mathscr{E}/1-\mathscr{E}')\, R_1(\mathscr{E}')\, d\mathscr{E}'$$

$$(4.17b) \quad L_2(s+1) = 4 \int_0^1 \mathscr{E}^{s-1}\, d\mathscr{E} \int_{\mathscr{E}}^1 N_1(\mathscr{E}/\mathscr{E}')\, R_2(\mathscr{E}')\, d\mathscr{E}'$$

$$+ 2 \int_0^{1/2} \mathscr{E}^{s-1}\, d\mathscr{E} \int_{\mathscr{E}}^{1-\mathscr{E}} [N_1(\mathscr{E}/\mathscr{E}') + M_1(\mathscr{E}/1-\mathscr{E}')$$

$$+ 1 + N_1(\mathscr{E}/\mathscr{E}')\, M_1(\mathscr{E}/\mathscr{E}')]\, R_2(\mathscr{E}')\, d\mathscr{E}'.$$

Thus $N_2(\mathscr{E})$ and $M_2(\mathscr{E})$ can be explicitly solved once $N_1(\mathscr{E})$ and $M_1(\mathscr{E})$ are explicitly obtained.

4.2. SOLUTION OF THE MEAN NUMBERS

We now turn to the evaluation of the contour integrals given by (4.10) and (4.11). The integrand has no singularities to the right of the line $s = 2$ in the complex s-plane. At $s = 2$, there is a simple pole due to the zero of $\phi(s) = A(s)\, D - B(s)\, C(s)$. An examination of equations (5.12) through (5.15) of Chapter 5 shows that there are possible poles at $s = 1, 0, -1,$ and -2 arising from the factor $B(s)\, C(s)$. In addition there are poles arising from the possible zeros of $\phi(s)$.

Let us first evaluate the integral on the right-hand side of (4.10). Since $A(s)$ is a linear combination of $\psi(s)$ the digamma function* and $1/s(s+1)$, $A(s)$ has poles at all negative integral values of s. Hence it is easy to see

$$\lim B(s)C(s)/\phi(s)$$

exists when s tends to any one of the points 0, -1, and -2. Thus the integrand has a simple pole at $s = 1$ due to the factor $(s-1)$ in the denominator and another simple pole at $s = 2$ due to the simple zero of $\phi(s)$. By consideration of signs of $B(s)\,C(s)$ and $A(s)$, it is easy to prove that $\phi(s)$ has no zeros on the portion of the real axis extending from -3 to 2. From -3 onward, there is a zero of $\phi(s)$ in the interval $(-n-1, -n)$, n being an integer, this being due to $\psi(s)$ taking all values from $-\infty$ to $+\infty$ in the interval $(-n-1, -n)$. Apart from this, there may be zeros of $\phi(s)$ off the real axis. However, it is shown in the next section that $\phi(s)$ has six zeros off the real axis.**

Choosing $\sigma = \sigma_0 > 2$ we can evaluate the Mellin integral. Let us consider a typical pole at $s = +\lambda_m$ on the negative real axis $(-m-1 < \lambda_m < -m)$. The residue R_m from λ_m is given by

$$(4.18) \qquad R_m = \lim_{s \to \lambda_m} (s - \lambda_m)\, \frac{B(s)\, C(s)\, e^{y(s-1)}}{\phi(s)\,(s-1)}.$$

Since

$$(4.19) \qquad \lim \frac{s - \lambda_m}{\phi(s)} = \lim \left\{ \frac{[\phi(s) - \phi(\lambda_m)]}{s - \lambda_m} \right\}^{-1} = \frac{1}{\phi'(\lambda_m)}$$

we find

$$(4.20) \quad R_m = \frac{e^{y(\lambda_m - 1)}}{(\lambda_m - 1)} \frac{B(\lambda_m)\, C(\lambda_m)}{D A'(\lambda_m) - B'(\lambda_m)\, C(\lambda_m) - B(\lambda_m)\, C'(\lambda_m)}.$$

We note that $|B(s)\, C(s)|$ is uniformly bounded in the strip $-m < s < -2$ and hence we obtain

$$(4.21) \qquad |R_m| < e^{y(\lambda_m - 1)} \frac{M}{|A'(\lambda_m) + \delta|}$$

where M and δ are some positive numbers less than unity. We next invoke the

---

* $\psi(s)$ is defined to be $d/ds\,(\log s!)$.
** For the sake of convenience, we have deferred the discussion to the final subsection.

following useful property of the digamma function

(4.22)
$$\left|\frac{10}{\psi'(s)}\right| \leqslant 1$$

for s real.

Using the above inequality, we notice

(4.23)
$$|R_m| < \frac{e^{y(\lambda_m - 1)}}{10}$$

and hence

(4.24)
$$\left|\sum_{i=1}^{\infty}\right| < \frac{e^{-y}}{10(1 - e^{-y})}.$$

Thus $N_1(\mathscr{E})$ can very well be approximated by the sum of the residues from the poles at $s = 1$ and 2, provided we can neglect the contribution from the six zeros of $\phi(s)$ lying off the real axis. We have not been able to establish mathematically that the contribution is negligible. However, the close agreement of the mean numbers arising from the poles on the real axis with those calculated numerically shows that the zeros of $\phi(s)$ may be in the half-plane: real $s < 0$. Under this approximation $N_1(\mathscr{E})$ is given by

(4.25) $N_1(\mathscr{E}) = \beta e^y - 1 = \beta/\mathscr{E} - 1,$ $\beta = 0.443.$

In an exactly similar way $M_1(\mathscr{E})$ can be approximated and we find

(4.26) $M(\mathscr{E}) = \beta/\mathscr{E}.$

In Table 16, we have presented the value of N_1 and M_1 as given by (4.25) and (4.26) as well as those computed earlier numerically. The agreement

Table 16

y	M_1	N_1	$N_1{}^*$
4	24.18	23.18	23.15
5	65.71	64.71	64.71
6	178.6	177.6	177.6
7	485.6	484.6	484.6
8	1320	1319	1319
9	3588	3587	3587
10	9753	9752	9752
11	2651×10	2651×10	2651×10
12	7206×10	7206×10	7206×10
13	1959×10^2	1959×10^2	1959×10^2
14	5325×10^2	5325×10^2	5325×10^2
15	1447×10^3	1447×10^3	1447×10^3

* Obtained by numerical evaluation of the inversion integral.

with the numerically computed values shows the dominant nature of the pole at $s = 2$.

4.3. HIGHER MOMENTS OF THE NUMBER DISTRIBUTION

Now we can solve for the second factorial moments from (4.15) and (4.16):

$$(4.27) \qquad N_2(\mathscr{E}) = \frac{1}{2\pi i} \int_{\sigma-i\infty}^{\sigma+i\infty} \frac{D\,L_1(s) + C(s)\,L_2(s)}{A(s)\,D - B(s)\,C(s)}\, e^{y(s-1)}\, ds$$

$$M_2(\mathscr{E}) = \frac{1}{2\pi i} \int_{\sigma-i\infty}^{\sigma+i\infty} \frac{B(s)\,L_1(s) + A(s)\,L_2(s)}{A(s)\,D - B(s)\,C(s)}\, e^{y(s-1)}\, ds.$$

$L_1(s)$ and $L_2(s)$ can be explicitly calculated

$$(4.28) \quad L_1(s) = \frac{(\frac{1}{2})^{s-3}}{s-3}\frac{\beta^2}{6}(1 + 4\delta) - \frac{(\frac{1}{2})^{s-2}}{s-2}2\beta\left(\beta\delta + \frac{2+3\delta}{6}\right)$$

$$+ \frac{(\frac{1}{2})^{s-1}}{s-1}2\beta\left(\beta\delta - \frac{\beta}{2} + 1 + \delta\right)$$

$$- \frac{(\frac{1}{2})^{s}}{s}2\beta\left(\beta\frac{2\delta-1}{3} + 1\right) + \frac{(\frac{1}{2})^{s+1}}{s+1}\frac{4}{3}\beta$$

$$L_2(s) = \beta^2\frac{5\delta-1}{15}\frac{(\frac{1}{2})^{s-3}}{s-3} - \frac{1}{s-1}[2B(s) + 2\beta^2(\frac{1}{2})^{s-1}]$$

$$+ 2\beta\frac{B(s)}{s-2} + \frac{4}{3}(1+\delta)\beta^2\frac{(\frac{1}{2})^{s}}{s} - 2\beta^2\delta\frac{(\frac{1}{2})^{s+1}}{s+1}$$

$$+ \frac{4\beta^2}{5}\delta\frac{(\frac{1}{2})^{s+2}}{s+2}, \qquad\qquad \delta = \frac{4}{3} + \alpha.$$

In view of the occurrence of $\phi(s)$ in the denominator of the integrand in (4.27) the arguments used in Section 3 are also equally applicable. $N_2(\mathscr{E})$ and $M_2(\mathscr{E})$ can be evaluated by the calculation of residues from the poles of the numerator in (4.27) and the dominant zero of $\phi(s)$ at $s = 2$. After some calculations, we obtain

$$(4.29) \qquad N_2(\mathscr{E}) = 0.2102\, e^{2y} + 0.2134\, e^{y} + 2.2534$$

$$M_2(\mathscr{E}) = 0.2420\, e^{2y} + 0.2134\, e^{y}.$$

The mean square number is obtained from $N_2(\mathscr{E})$ and $M_2(\mathscr{E})$ by adding $N_1(\mathscr{E})$ and $M_1(\mathscr{E})$ respectively. In cascade theory, it is customary to compare the mean square deviation with the deviation corresponding to Poisson and

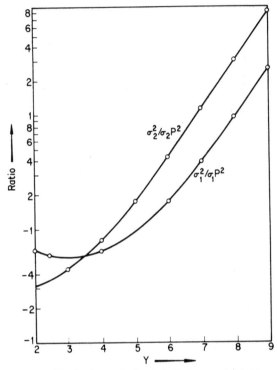

Fig. 7. Variation of σ^2 with Poisson deviation.

Furry distribution. Figures 7 and 8 demonstrate the variation of σ_1^2/σ_{1P}^2, σ_2^2/σ_{2P}^2, σ_1^2/σ_{1F}^2, and σ_2^2/σ_{2F}^2 with y.

It is interesting to note that we can obtain successively all the factorial moments of the number distribution by successive differentiation of (4.5) and (4.6) at $u = 1$. To illustrate the possibility, we have calculated the third factorial moments of the number distribution. It is obvious from (4.5) and (4.6) that the equation will contain the second and first factorial moments whose values are given by (4.25), (4.26), and (4.29). Since the method of calculation is straightforward, we give below the final expressions for $N_3(\mathscr{E})$ and $M_3(\mathscr{E})$:

(4.30) $\quad N_3(\mathscr{E}) = 0.2802\,e^{3y} + 0.2178\,e^{2y} - 0.6430\,e^{y} + 2.0383$

$\quad\quad\quad M_3(\mathscr{E}) = 0.1921\,e^{3y} + 0.8659\,e^{2y} - 0.6430\,e^{y}.$

The possibility of explicit solutions of the moments for the special case of infinite thickness gives us the hope that it may be possible to obtain an explicit solution for cascades of finite thicknesses.

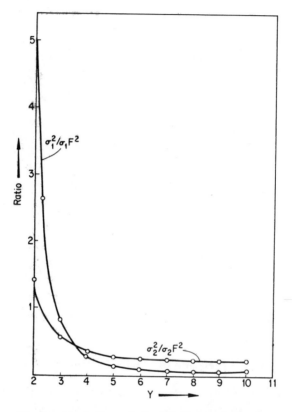

Fig. 8.   Variation of σ^2 with Furry deviation.

4.4.   ZEROS OF $\phi(s)$

We shall examine the zeros of $\phi(s)$ lying off the real axis in the complex s-plane. We recall that $\psi(s)$ has the asymptotic expression for $|s| \gg 1$, $\arg s \neq \pm \pi$:

$$(4.31) \qquad \psi(s) = \log s + \frac{1}{2s} - \frac{1}{12s^2} + \frac{1}{120s^4}.$$

Thus we rewrite $\phi(s)$ as

(4.32) $\phi(s) = \left(\dfrac{4}{3} + \alpha\right) D\psi(s) - 0.0497 - \dfrac{D}{s(s+1)} - B(s)\,C(s)$

$\qquad\qquad = g(s) + h(s)$

where

(4.33) $$g(s) = \left(\dfrac{4}{3} + \alpha\right) D\psi(s).$$

With the help of (4.31), it is easy to see that

$$|g(s)| > |h(s)|$$

for

$$|s| \gg 1 \qquad \text{and} \qquad \text{Arg } s \neq \pm\,\pi.$$

Let us consider the contour shown in Figure 9. A is the point whose s

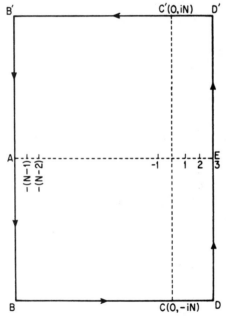

Fig. 9.   Contour chosen for the determination of the zeros of $\phi(s)$.

value is $-N + \delta$ where δ is a small positive number. We shall choose N arbitrarily large; E is the point $(3, 0)$ in the Argand s-plane. Under these conditions, it is easy to verify that

$$|g(s)| > |h(s)|$$

on the contour $ABCDED'C'B'A$. Since $g(s)$ and $h(s)$ are meromorphic functions of s, it follows by Roche's theorem (see, for example, Einar Hille [29]) that the difference between the number of zeros and poles of $g(s)$ included in the region enclosed by the contour shown in Figure 9 is the same as that of $\phi(s)$. From the properties of the digamma function, it is clear that the difference is one. However, on the portion of the real axis included in the contour, there are poles at negative integral values of s as well as at $s = 0$ and 1. All these poles are simple except at $s = 0$ and -1 which are second order poles for $\phi(s)$. In addition, $\phi(s)$ has simple zeros in each of the intervals $(-m, -m + 1)$, m being >3, there being no further zeros in the interval $(-3, 2)$ except at the point 2 where $\phi(s)$ has a simple zero. Thus the difference between the number of zeros and poles of $\phi(s)$ is -5. Since the difference is $+1$ for $g(s)$, $\phi(s)$ has *six* zeros off the real axis. From the asymptotic properties of $\psi(s)$, it can be shown that the zeros should lie in a region limited by the circle $|z| = r$, r being of the order of unity. However, the agreement of the mean numbers obtained by neglecting the contributions from these zeros with those obtained by saddle point method as well as numerical contour integration shows that the zeros should have negative real parts.

5. Sequent Correlations in Cascades

We have already seen the utility of the sequent correlation densities in the description of electron-photon showers. The solution for the sequent correlation density of degree one has been obtained in Section 8 of the previous chapter. These solutions acquire a special importance in view of some of the recent experimental findings of Zaimidoroga *et al.* [30] who have observed the correlation of the number of electrons corresponding to two different thicknesses. We shall present some of the available analytical and numerical results pertaining to such sequent correlations [31]. If we define $N(E, E_0, t_1, t)$ to be the mean number of electrons that are produced between 0 and t_1 having an energy greater than E at $t > t_1$ (E_0 being the energy of a primary electron), we have

$$(5.1) \qquad N(E, E_0, t_1, t) = \frac{1}{2\pi i} \int_{\sigma - i\infty}^{\sigma + i\infty} \frac{\mathscr{P}_1^1(1, t_1, s_2, t)}{s - 1} \left(\frac{E_0}{E}\right)^{s-1} ds$$

where $\mathscr{P}_1^1$ is given by equation (7.11a) of the previous chapter.

It is rather difficult to invert the above expression analytically. However, it can be calculated by the method of steepest descents. Table 17 gives the values of $N(E/E_0, t_1, t)$ for $t_1 = 1$ and various values of t and $y = \log(E_0/E)$.

The table shows that most of the electrons that are produced between 0 and 1 drop down considerably in energy by the time they traverse 3 or 4 cascade units. Such a table can be used in conjunction with the study of

Table 17

$N(E/E_0, t_1, t)$; $t_1 = 1$

y	3	4	t_2 5	6	7
3	0.1408	0.0637	0.0108	0.0023	0.0004
4	0.3763	0.1295	0.0453	0.0130	0.0082
5	0.5340	0.3386	0.0857	0.0445	0.0139
6	0.4253	0.5239	0.2927	0.0954	0.0133
7	0.2678	0.2112	0.5182	0.2575	0.0462

"young" and "old" electrons that are observed at any particular thickness.

We next deal with $\mathscr{F}_2^i(E_1; t_1; E_2, t_2; E_3, E_4, t|E_0)$ the sequent correlation density of degree two of electrons produced between t_1 and $t_1 + dt_1$ and t_2 and $t_2 + dt_2$ and observed at a later thickness $t > t_1, t_2$. Exactly as in the first part of the section, we analyze the possible outcome of events in $(0, \Delta)$ of the t-axis and use the invariant imbedding technique. Thus we have

(5.2) $\mathscr{F}_2^i(E_1, t_1; E_2, t_2; E_3, E_4, t|E_0)$

$= (1 - \Delta \int R_i(E'|E_0)\, dE')\, \mathscr{F}_2^i(E_1, t_1 - \Delta; E_2, t_2 - \Delta; E_3, E_4, t - \Delta|E_0)$

$+ \Delta \int R_i(E'|E_0)[\mathscr{F}_2^1(E_1, t_1 - \Delta; E_2, t_2 - \Delta; E_3, E_4, t - \Delta|E')$

$\quad + \mathscr{F}_2^{3-i}(E_1, t_1 - \Delta; E_2, t_2 - \Delta; E_3, E_4, t - \Delta|E_0 - E')$

$\quad + \mathscr{F}_1^1(E_1, t_1 - \Delta; E_3, t - \Delta|E')$
$\qquad\qquad \mathscr{F}_1^{3-i}(E_2, t_2 - \Delta; E_4, t - \Delta|E_0 - E')$

$\quad + \mathscr{F}_1^{3-i}(E_1, t_1 - \Delta; E_3, t - \Delta|E_0 - E')$
$\qquad\qquad \mathscr{F}_1^1(E_2, t_2 - \Delta; E_4, t - \Delta|E')]\, dE'.$

By letting $\Delta \to 0$, we obtain

(5.3) $\left(\dfrac{\partial}{\partial t_1} + \dfrac{\partial}{\partial t_2} + \dfrac{\partial}{\partial t}\right) \mathscr{F}_2^i(E_1, t_1; E_2, t_2; E_3, E_4, t|E_0)$

$= -\int R_i(E'|E_0)[\mathscr{F}_2^i(E_1, t_1; E_2, t_2; E_3, E_4, t|E_0)$

$\qquad\qquad -\mathscr{F}_2^1(E_1, t_1; E_2, t_2, E_3, E_4, t|E'\, E')]\, dE'$

$+ \int R_i(E'|E_0)[\mathscr{F}_2^{3-i}(E_1, t_1; E_2, t_2; E_3, E_4, t|E_0 - E')$

$\qquad\qquad + \mathscr{F}_1^1(E_1, t_1; E_3, t|E')\, \mathscr{F}_1^{3-i}(E_2, t_2; E_4, t|E_0 - E')$

$\qquad\qquad + \mathscr{F}_1^{3-i}(E_1, t_1; E_3, t|E_0 - E')\, \mathscr{F}_1^1(E_2, t_2; E_4, t|E')]\, dE'.$

The boundary conditions satisfied by $\mathscr{F}_2^i$ are given by

(5.4a) $$\mathscr{F}_2^1(E_1, 0; E_2, 0; E_3, E_4, t|E_0) = 0$$

(5.4b) $\mathscr{F}_2^2(E_1, 0; E_2, 0; E_3, E_4, t|E_0)$
$$= 2 R_2(E_1|E_0)\, \delta(E_0 - E_1 - E_2)\, \pi(E_3|E_1; t)\, \pi(E_4|E_2; t)$$

(5.4c) $$\mathscr{F}_2^1(E_1, 0; E_2, t_2; E_3, E_4, t|E_0) = 0$$

(5.4d) $$\mathscr{F}_2^1(E_1, t_1; E_2, 0; E_3, E_4, t|E_0) = 0.$$

In addition to the above conditions, we also need the value of

$$\mathscr{F}_2^2(E_1, 0, E_2, t_2, E_3, E_4, t|E_0) \quad \text{and} \quad \mathscr{F}_2^2(E_1, t_1, E_2, 0; E_3, E_4, t|E_0).$$

These cannot be specified directly but can be obtained after some calculations which are straightforward. However we observe that by definition

(5.5) $\mathscr{F}_2^2(E_1, 0; E_2, t_2; E_3, E_4, t|E_0) = R_2(E_1|E_0)\, [\mathscr{F}\pi_1^2(E_2, t_2; E_3, E_4, t|E_1)$
$$+ \pi(E_3|E_1; t_1)\, \mathscr{F}_1^2(E_2, t_2; E_3, t|E_0 - E_1)]$$

where $\mathscr{F}\pi_1^2(E_2, t_2; E_3, E_4, t|E_1)\, dE_2\, dE_3\, dE_4\, dt_2$ denotes the joint probability that an electron of primitive energy between E_2 and $E_2 + dE_2$ is created between t_2 and $t_2 + dt_2$ in the shower excited by an electron having an energy E_1, the primary and the secondary electrons so created dropping to energies lying in the intervals $(E_3, E_3 + dE_3)$ and $(E_4, E_4 + dE_4)$ respectively. $\mathscr{F}\pi_1^2$ is an unknown function; however, this need not trouble us since it is easy to write down the integral equation satisfied by $\mathscr{F}\pi_1^2$, by the use of invariant imbedding techniques. $\mathscr{F}\pi_1^2$ by itself is important, in the study of bursts produced by mu-mesons and in interpreting portions of showers accompanied or not by the primary. We next proceed to obtain $\mathscr{F}\pi_1^i$ explicitly. To obtain the differential equation satisfied by this function, we consider the first infinitesimal interval $(0, \Delta)$ of thickness t. In $(0, \Delta)$ the primary electron may radiate a photon and drop to a lower energy in which case we have two independent primaries (electron and photon). If the primary electron does not radiate, the contribution to the density function is the density itself except that the thicknesses are reduced by Δ. Thus by simple probability arguments, we obtain

(5.6) $$\left(\frac{\partial}{\partial t_1} + \frac{\partial}{\partial t}\right) \mathscr{F}\pi_1^1(E_1, t_1; E_2, E_3, t|E_0)$$

$$= -\int_0^{E_0} R_1(E'|E_0)\, \mathscr{F}\pi_1^1(E_1, t_1; E_2, E_3, t|E_0)\, dE'$$

$$+ \int_0^{E_0} R_1(E'|E_0)\ \mathscr{F}\pi_1^1(E_1, t_1; E_2, E_3, t|E')\ dE'$$

$$+ \int_0^{E_0} R_1(E'|E_0)\ \mathscr{F}_1^2(E_1, t_1; E_2, E_3, t|E_0 - E')\,\pi(E_3, t|E')\ dE'.$$

We note that $\mathscr{F}\pi_1^1$ is a function only of $\mathscr{E}_1 = E_1/E_0$, $\mathscr{E}_2 = E_2/E_0$, $\mathscr{E}_3 = E_3/E_0$ and hence we write it as $\mathscr{F}\pi_1^1(\mathscr{E}_1, t_1; \mathscr{E}_2, \mathscr{E}_3, t)$. Defining the Mellin transform of $\mathscr{F}\pi_1^1$ with respect to $\mathscr{E}_1, \mathscr{E}_2$, and $\mathscr{E}_3$ as $\mathscr{F}\pi_1^1(s_1, t_1; s_2, s_3, t)$ we obtain

$$(5.7) \qquad \left(\frac{\partial}{\partial t_1} + \frac{\partial}{\partial t}\right) \mathscr{F}\pi_1^1(s_1, t_1; s_2, s_3, t)$$

$$= -A(s_1 + s_2 + s_3 - 2)\ \mathscr{F}\pi_1^1(s_1, t_1; s_2, s_3, t)$$

$$+ \mathscr{F}_1^2(s_1, t_1; s_2, t)\,\pi(s_3, t)\,a_2(s_1 + s_2 - 1, s_3).$$

Equation (5.7) can be solved by taking a double Laplace transformation with respect to t_1 and t and using the solutions for $\mathscr{F}_1^2(s_1, t_1; s_2, t)$ and $\pi(s_3, t)$. After some calculation we obtain

$$(5.8) \quad \mathscr{F}\pi_1^1(s_1, t_1; s_2, s_3, t) = \frac{a_1(s_1 + s_2 - 1, s_3)\,B(s_1 + s_2 - 1)}{\mu(s_1 + s_2 - 1) - \lambda(s_1 + s_2 - 1)} \times$$

$$[\phi(t_1)\,e^{-A(s_2)t - A(s_3)t} - \phi(t_1 - t)\,e^{-A(s_1 + s_2 + s_3 - 2)t}]$$

where

$$(5.9) \qquad \phi(t_1) = \frac{\mu(s_1 + s_2 - 1) - D}{A(s_1 + s_2 + s_3 - 2) - A(s_3) - \lambda(s_1 + s_2 - 1)} \times$$

$$(\exp\{-[\lambda(s_1 + s_2 - 1) - A(s_2)]t_1\}$$

$$- \exp\{-[A(s_1 + s_2 + s_3 - 2) - A(s_2) - A(s_3)]t_1\})$$

$$- \frac{D - \lambda(s_1 + s_2 - 1)}{A(s_1 + s_2 + s_3 - 2) - A(s_3) - \mu(s_1 + s_2 - 1)} \times$$

$$(\exp\{-[\mu(s_1 + s_2 - 1) - A(s_2)]t_1\}$$

$$- \exp\{-[A(s_1 + s_2 + s_3 - 2) - A(s_2) - A(s_3)]t_1\}).$$

The mean number of electrons that are produced between 0 and t, the electrons and the primary remaining above an energy E at t, is given by

$$(5.10) \quad N(E, E_0, t) = \frac{1}{(2\pi i)^2} \int_{\sigma-i\infty}^{\sigma+i\infty} \int_{\sigma-i\infty}^{\sigma+i\infty} \int_{0}^{t} \mathscr{F}\pi_1^1(1, t_1, s_2, s_3, t) \\ e^{y(s_2+s_3-2)} \, ds_2 \, ds_3 \, dt_1$$

where $y = \log(E_0/E)$.

Equation (5.10) can be numerically evaluated by the saddle point method.

6. Bursts Produced by Muons and Electrons

The fluctuations in the size of showers produced by mu-mesons (muons) has received considerable importance in the interpretation of cosmic ray events. Particularly in view of the occurrence of highly energetic muons and photons, it has become necessary to study the average size of the soft showers as well as their fluctuations about the mean size. Recently large bursts of showers excited by muons and electrons have been reported by Christy and Kusaka [32]. Mitra [33] has attempted to estimate the mean size and the fluctuation about the mean on the basis of a Polya distribution. In this section, we shall calculate the mean and the mean square of the number of the electrons on the basis of the new approach to the cascade theory.

6.1. BASIC EQUATIONS

Let us consider the shower excited by a muon of energy E_0 incident at the top of the atmosphere corresponding to $t = 0$. We assume that the shower develops due to the following fundamental processes:

(i) Radiation of a photon by the primary muon;
(ii) Radiation of a photon by a secondary electron; and
(iii) Production of an electron-positron pair by a secondary photon.

When screening is complete, the probability that a mu-meson of energy E to radiate a quantum and drop down to an energy between E' and $E' + dE'$ is given by

$$(6.1) \quad R_\mu(E'|E) \, dE' = \frac{1}{(210)^2} R_1(E'|E) \, dE'$$

where $R_1(E'|E)$ is given by (2.1) of Chapter 5. Let $\pi_i(n, E, E_0, t)$ be the probability of finding n electrons between 0 and t in a shower excited by a primary of i-th type ($i = 1, 2, 3$ denote respectively a muon, a photon, and an electron) each of the electrons having an energy greater than E at the time of its production. In the first approximation we shall neglect the loss in energy of muons and electrons due to ionization. Then it is easy to see that $\pi_i(n, E, E_0, t)$ is a function of $\mathscr{E} = E/E_0$, due to the homogeneous nature of the assumed cross-sections. To obtain the equations satisfied by $\pi_i(n, \mathscr{E}, t)$ we use the technique of invariant imbedding as demonstrated in the previous chapter:

$$(6.2) \qquad \frac{\partial \pi_1(n, \mathscr{E}, t)}{\partial t} = -\pi_1(n, \mathscr{E}, t) \int_0^1 R_\mu(\mathscr{E}') \, d\mathscr{E}'$$

$$+ \sum_{n_1+n_2=n} \int_0^1 R_\mu(\mathscr{E}') \, \pi_1(n, \mathscr{E}/\mathscr{E}', t) \, \pi_2(n, \mathscr{E}/1 - \mathscr{E}', t) \, d\mathscr{E}'$$

$$(6.3) \qquad \frac{\partial \pi_2(n, \mathscr{E}, t)}{\partial t} = -\pi_2(n, \mathscr{E}, t) \int_0^1 R_2(\mathscr{E}') \, d\mathscr{E}'$$

$$+ \int_0^\mathscr{E} R_2(\mathscr{E}') \, \pi_3(n - 1, \mathscr{E}/1 - \mathscr{E}', t) \, d\mathscr{E}'$$

$$+ \sum_{m=0}^\infty \int_\mathscr{E}^{1-\mathscr{E}} R_2(\mathscr{E}') \, \pi_3(n, \mathscr{E}/\mathscr{E}', t) \, \pi_3(n - m - 2, \mathscr{E}/1 - \mathscr{E}', t) \, d\mathscr{E}'$$

$$+ \int_{1-\mathscr{E}}^1 R_2(\mathscr{E}') \, \pi_3(n - 1, \mathscr{E}/\mathscr{E}', t) \, d\mathscr{E}'$$

$$(6.4) \qquad \frac{\partial \pi_3(n, \mathscr{E}, t)}{\partial t} = -\pi_3(n, \mathscr{E}, t) \int_0^1 R_2(\mathscr{E}') \, d\mathscr{E}'$$

$$+ \sum_{n_1+n_2=n} \int_0^1 R_2(\mathscr{E}') \, \pi_3(n_1, \mathscr{E}/\mathscr{E}', t) \, \pi_2(n_2, \mathscr{E}/1 - \mathscr{E}', t) \, d\mathscr{E}'.$$

We observe that (6.2) and (6.4) are valid for the entire range of $\mathscr{E}$ while (6.3) is valid only for $0 \leqslant \mathscr{E} \leqslant \frac{1}{2}$. For $\mathscr{E} > \frac{1}{2}$, π_2 satisfies the equation

$$(6.3a) \qquad \frac{\partial \pi_2(n, \mathscr{E}, t)}{\partial t} = -\pi_2(n, \mathscr{E}, t) \int_0^t R_2(\mathscr{E}') \, d\mathscr{E}'$$

$$+ \int\limits_0^{1-\mathscr{E}} R_2(\mathscr{E}')\, \pi_3(n-1, \mathscr{E}/1-\mathscr{E}', t)\, d\mathscr{E}'$$

$$+ \int\limits_{\mathscr{E}}^1 R_2(\mathscr{E}')\, \pi_3(n-1, \mathscr{E}/\mathscr{E}', t)\, d\mathscr{E}'.$$

It is interesting to note that in the case of a shower generated by an electron or a photon, (6.3) and (6.4) alone need be considered.

Defining the generating function $g_i(u, \mathscr{E}, t)$ as

$$(6.5) \qquad\qquad g_i(u, \mathscr{E}, t) = \sum_0^\infty u^n\, \pi_i(n, \mathscr{E}, t)$$

and observing that the cumulant moments are given by

$$(6.6) \qquad\qquad n_i^m(\mathscr{E}, t) = \frac{\partial^m}{\partial u^m}\, g_i(u, \mathscr{E}, t)\Big|_{u=1}$$

we find

$$(6.7) \qquad \frac{\partial n_1^m(\mathscr{E}, t)}{\partial t} = -n_1^m(\mathscr{E}, t) \int\limits_0^1 R_\mu(\mathscr{E}')\, d\mathscr{E}'$$

$$+ \sum_{i=0}^m \int\limits_0^1 R_\mu(\mathscr{E}')\, n_1^i(\mathscr{E}/\mathscr{E}', t)\, n_2^{m-i}(\mathscr{E}/1-\mathscr{E}', t)\, d\mathscr{E}'.$$

We do not propose to go into the similar equations satisfied by $n_2^m(\mathscr{E}, t)$ and $n_3^m(\mathscr{E}, t)$ since they have been dealt with in general detail in the earlier section.

Equation (6.7) can be rewritten as

$$(6.8) \qquad \frac{\partial n_1^m(\mathscr{E}, t)}{\partial t} = - \int\limits_0^1 R_\mu(\mathscr{E}')\, [n_1^m(\mathscr{E}, t) - n_1^m(\mathscr{E}/\mathscr{E}', t)]\, d\mathscr{E}'$$

$$+ \int\limits_0^1 R_\mu(\mathscr{E}')\, n_2^m(\mathscr{E}/1-\mathscr{E}', t)\, d\mathscr{E}'$$

$$+ \sum_{i=1}^{m-1} \int\limits_0^1 R_\mu(\mathscr{E}')\, n_1^i(\mathscr{E}/\mathscr{E}', t)\, n_2^{m-i}(\mathscr{E}/1-\mathscr{E}', t)\, d\mathscr{E}'.$$

Equation (6.8) can be solved recursively if we know $n_2^i(\mathscr{E}, t)$. These are just the cumulant moments of the number distribution corresponding to a photon primary.

The Mellin transform solution of $n_1^1(\mathscr{E}, t)$ can be obtained without any difficulty.* Using the well-known expressions for $n_2^i(s, t)$ we obtain, after straightforward calculations,

$$(6.9) \quad n_1^1(s, t) = \frac{B(s + 1)\, C^\mu(s + 1)}{s}\left[\frac{\mu(s + 1) - D}{\lambda(s + 1)}\left\{\frac{1 - e^{-A^\mu(s+1)t}}{A^\mu(s + 1)}\right.\right.$$

$$-\left.\frac{e^{-\lambda(s+1)t} - e^{-A^\mu(s+1)t}}{A^\mu(s + 1) - \lambda(s + 1)}\right\} + \frac{D - \lambda(s + 1)}{\mu(s + 1)}\left\{\frac{1 - e^{-A^\mu(s+1)t}}{A^\mu(s + 1)}\right.$$

$$\left.\left.-\frac{e^{-\mu(s+1)t} - e^{-A^\mu(s+1)t}}{A^\mu(s + 1) - \mu(s + 1)}\right\}\right]$$

where the functions A, B, C, D are the same as those defined in Section 5 of Chapter 5 and the functions $A^\mu(s + 1)$ and $C^\mu(s + 1)$ are the corresponding functions associated with the muon cross-sections.

6.2. SHOWERS CORRESPONDING TO LARGE THICKNESSES

We next consider an interesting case corresponding to large t. Proceeding to the limit as t tends to infinity, we find

$$(6.10) \quad n_1^1(s) = \frac{B(s + 1)\, C^\mu(s + 1)\, A(s + 1)}{A^\mu(s + 1)\, s\, [A(s + 1)D - B(s + 1)\, C(s + 1)]}.$$

If we observe that $A^\mu(s + 1)$ and $C^\mu(s + 1)$ differ from $A(s + 1)$ and $C(s + 1)$ only by a constant factor arising from the mass, equation (6.10) can be identified to be the solution corresponding to an electron primary. This should be expected since we have made t infinite. In fact, the result is true for all the moments of number distribution. Thus the muon characteristics of a shower are completely lost by making t tend to infinity. However, we can introduce the decay constant of the mu-meson and this has a tendency to cut the size of the shower. The decay constant can be introduced into the equations.

* We use the same functional symbol n_1 to denote the moment as well as its Mellin transform, the distinction being apparent from the context.

We know that in a muon shower the decay probability per unit thickness is given by

$$k = \frac{mc}{\tau E}$$

where m is the rest energy, τ the lifetime of the muon. In this expression E is the only varying quantity. For large values of E_0 it is usually assumed (see, for example, reference [33]) that the change in energy of the muon is small compared to the initial muon energy. Thus in the expression for k, E is replaced by E_0. However, this is not a consistent procedure since muons may lose a considerable part of the energy while traversing a very thick layer of matter. We shall estimate the correction on this assumption and compare it with the results obtained by taking the exact expression for k. Thus the effect of the decay constant can be incorporated by redefining $A^\mu(s+1)$ as

$$(6.11) \qquad A^\mu(s+1) = k + \frac{1}{(210)^2} A(s+1)$$

where

$$(6.12) \qquad k = \frac{mc}{\tau E_0}.$$

$C^\mu(s+1)$ remains unaltered and is equal to $(210)^{-2} C(s+1)$. Since k is very small, $n_1^1(s)$ can be estimated easily. Retaining only the first power of k, we obtain

$$(6.13) \qquad n_1^1(s) = \frac{B(s+1)\,C(s+1)}{s[A(s+1)\,D - B(s+1)\,C(s+1)]}\left[1 - \frac{(210)^2}{A(s+1)}\right].$$

Inverting the above expression, we find

$$(6.14) \qquad n_1^1(\mathscr{E}) = 0.443\, e^y - (210)^2\, k\, (0.4377\, e^y - 0.335\, y).$$

For the primary muon energy of the order of 1000 GeV (BeV), the correction is of the order of $10^{-4}\, e^y$.

We next use the exact expression for k. In this case, we obtain the difference equation for the mean number:

$$(6.15) \qquad A(s+1)\, n_1^1(s) = C(s+1)\, n_2^1(s) - (210)^2\, k\, n_1^1(s-1)$$
$$= C(s+1)\, n_2^1(s) - (210)^2\, k\, \delta n_1^1(s)$$

where δ is a unit difference operator. Thus

$$(6.16) \qquad [A(s+1) + (210)^2\, k\delta]\, n_1^1(s) = C(s+1)\, n_2^1(s).$$

Since $(210)^2 k$ is small for fairly large muon primary energies, we can solve the difference equation to first order in $(210)^2 k$:

(6.17) $$n_1^1(s) = n_3^1(s) - \frac{(210)^2 k B(s) C(s) A(s)}{A(s+1)(s-1)[A(s)D - B(s)C(s)]}.$$

Inverting the above expression, we find

(6.18) $$n_1^1(\mathscr{E}) = 0.443 \, e^y - (210)^2 k(0.2897 \, e^{2y}).$$

Again for muon primary energy of the order of 1000 GeV, the correction is of the order of $10^{-4} \, e^{2y}$. Thus (6.18) and (6.17) differ by a factor e^y and except for small values of y, the difference between the two expressions is fairly large.

Corrections to the higher moments of the number distribution can easily be obtained. Using the same methods as were employed in the previous section, we can calculate the correction to the mean square number. The mean square number of electrons produced in the entire shower is given by

(6.19) $$\overline{[n(\mathscr{E})]^2} = 0.2102 \, e^{2y} + 0.656 \, e^y + 1.2534$$
$$-(210)^2 k \{0.017 \, e^{3y} + 0.41 \, e^{2y} + 0.3 \, e^y + 0.225 \, y\}.$$

Here again the corrections are small for fairly small values of y.

7. Cascade Theory under Approximation B

So far we have neglected the loss of energy suffered by electrons due to their characteristic property of ionizing the atoms in the medium. The loss in energy, although negligible at very high energies, cannot be neglected when the electrons drop down to moderate and low energies. In this section, we shall assume that the rate of loss of energy is a constant equal to β and that the bremsstrahlung and pair production phenomena are governed by the full screened cross-sections given by equations (2.1) and (2.2) of Chapter 5. In this case in the equations satisfied by the first order product densities, we have to add the term

$$\beta \frac{\partial f_1(E, t)}{\partial E}$$

to the right-hand side of the corresponding equation governing the electrons. The extra term introduces enormous difficulty in the explicit solution of the cascade equations. The Mellin transform solutions of the equations have been explicitly found by Bhabha and Chakrabarti [17] and by Snyder [34].

In this section, we shall deal with the solutions for the corresponding functions in the new approach to cascade theory.

7.1. MEAN BEHAVIOR OF SHOWERS IN (0, *t*)

If $F_1(E|E_0, t)$ is the product density of degree one of electrons produced between t and $t + dt$, we have (see reference [16])

(7.1) $$F_1(E|E_0; t) = 2\int g(E'|E_0; t) R_2(E|E') \, dE'.$$

The above equation is obtained by the simple observation that an electron can be produced only by a photon and as such the contribution can arise only from the product density of degree one of photons at t. Equation (7.1) is true whether or not we include ionization loss, the dependence on ionization loss entering through the photon instant product density. Tables of mean numbers on the basis of our new approach have been prepared by Ivanenko [35]. Ivanenko has observed that the mean number in the new approach should not be compared with the mean number of electrons that exist at t but with the functions obtained by integrating the instant product density over t. We do not share this view since such a function does not possess any probabilistic interpretation in general. However, in this particular case, the instant product densities continue to have some kind of a probabilistic interpretation even after integration over the variable t. To obtain such an interpretation we note that $f_1(E, t) \, dt$ represents the mean number of electrons each of which traverses the infinitesimal interval of thickness $(t, t + dt)$ with an energy equal to E. Then $f_1(E, t)$ for fixed E can be interpreted to be a product density over t-space since any particular electron can cross any thickness with the same energy only once with non-zero probability. Thus $f_1(E, t)$ can be considered as a product density either over E-space or over t-space but *not in the product space of E and t. However we wish to emphasize that the product density interpretation in t-space is not possible if there is a mechanism by which particles can gain energy, in which case any particle can cross several points on the t-axis with the same energy with non-zero probability.*

In view of the product density interpretation over t-space, we have the following interesting result: the mean number of electrons that are produced in the entire shower, the primitive energy of each of which is greater than E, is equal to $\int_E^\infty dE' \int_0^\infty f_1(E', t) \, dt$. This is due to the fact that the electrons that are produced with primitive energies greater than E have had an energy

$> E$ at some point on the t-axis with probability one. This interesting result has been observed and checked numerically by Ivanenko [35].

7.2. MOMENTS OF THE TOTAL NUMBER OF ELECTRONS PRODUCED

Using the notation of Section 4 of this chapter, we define $\pi_i(n, E, E_0, t)$ as the probability that n electrons above an energy E are produced between 0 and t by a primary of i-th type of energy E_0 ($i = 1$, $i = 2$ refer respectively to an electron and a photon). We shall assume that in addition to bremsstrahlung by electrons and pair creation by photons, electrons lose energy deterministically at a constant rate β per unit thickness of matter. In view of this assumption, $\pi_i(n, E, E_0, t)$ is no longer a function of E/E_0. However, this does not preclude using the method of Janossy [4] in deriving the integro-differential equations satisfied by $\pi_i(n, E, E_0; t)$. Using the same arguments as in Section 3 of Chapter 5, we obtain

$$(7.2) \quad \frac{\partial \pi_1(n, E, E_0, t)}{\partial t}$$

$$= -\pi_1(n, E, E_0, t) \int_0^{E_0} R_1(E'|E_0)\, dE'$$

$$+ \sum_{m=0}^{\infty} \int_0^{E_0} R_1(E'|E_0)\, \pi_1(m, E, E', t)\, \pi_2(n - m, E, E_0 - E', t)\, dE'$$

$$- \beta \frac{\partial \pi_1(n, E, E_0, t)}{\partial E_0}.$$

On the other hand, π_2 satisfies the same equation as the one corresponding to "no ionization loss." Setting $\lim_{t \to \infty} \pi_i(n, E, E_0, t) = \pi_i(n, E, E_0)$ we find

$$(7.3a) \quad \pi_1(n, E, E_0) \int_0^{E_0} R_1(E'|E_0)\, dE'$$

$$= \sum_{m=0}^{\infty} \int_0^{E_0} R_1(E'|E_0)\, \pi_1(m, E, E')\, \pi_2(n - m, E, E_0 - E')\, dE' - \beta \frac{\partial \pi_1(n, E, E_0)}{\partial E_0}$$

$$(7.3\text{b}) \quad \pi_2(n, E, E_0) \int_0^{E_0} R_2(E'|E_0) \, dE'$$

$$= \int_0^{E} R_2(E'|E_0) \, \pi_1(n - 1, E, E_0 - E') \, dE'$$

$$+ \sum_{m=}^{\infty} \int_{E_0}^{E_0 - E} R_2(E'|E_0) \, \pi_1(m, E, E') \, \pi_1(n - m - 2, E, E_0 - E') \, dE'$$

$$+ \int_{E_0 - E}^{E_0} R_2(E'|E_0) \, \pi_1(m - 1, E, E') \, dE', \, 0 < E \leqslant \tfrac{1}{2} E_0$$

$$(7.3\text{b}') \quad \pi_2(n, E, E_0) \int_0^{E_0} R_2(E'|E_0) \, dE' = \delta_{n0} \int_{E_0 - E}^{E_0} R_2(E'|E_0) \, dE'$$

$$+ \int_0^{E_0 - E} \pi_1(n - 1, E, E_0 - E') \, R_2(E'|E_0) \, dE'$$

$$+ \int_{E}^{E_0} R_2(E'|E) \, \pi_1(n - 1, E, E') \, dE', \, \tfrac{1}{2} E_0 < E \leqslant E_0.$$

Defining

$$(7.4) \qquad\qquad g_i(u, E, E_0) = \Sigma u^n \, \pi_i(n, E, E_0)$$

we obtain

$$(7.5) \quad g_i(u, E, E_0) \int_0^{E_0} R_i(E'|E_0) \, dE' = u^{i-1} \int_0^{E_0} g_{3-i}(u, E, E_0 - E') R_i(E'|E_0) \, dE'$$

$$+ u^{2i-2} \int_{E}^{E_0 - E} g_1(u, E, E') \, g_{3-i}(u, E, E_0 - E') \, R_i(E'|E_0) \, dE'$$

$$+ u^{i-1} \int_{E_0 - E}^{E_0} g_{3-i}(u, E, E') \, R_i(E'|E_0) \, dE' - \delta_{i1} \, \beta \, \frac{\partial g_1(u, E, E_0)}{\partial E_0}.$$

Equation (7.5) is valid for the entire range of E for $i = 1$ only. When $i = 2$, equation (7.5) covers only the range $0 < E \leqslant \frac{1}{2} E_0$. For $E > \frac{1}{2} E_0$, $g_2(u, E, E_0)$ satisfies the equation

$$(7.6) \quad g_2(u, E, E_0) \int_0^{E_0} R_2(E'|E_0) \, dE' = \int_{E_0-E}^{E} R_2(E'|E_0) \, dE'$$

$$+ u \int_0^{E_0-E} R_2(E'|E_0) \, g_1(u, E, E_0 - E') \, dE'$$

$$+ u \int_E^{E_0} R_2(E'|E_0) \, g_1(u, E, E') \, dE'.$$

Following the same notation as in Section 3, we define $N_m(E, E_0)$ and $M_m (E, E_0)$ as the m-th factorial moments. $N_1 (E, E_0)$ and $M_1 (E, E_0)$ satisfy the equations

$$(7.7) \quad N_1(E, E_0) \int_0^{E_0} R_1(E'|E_0) \, dE' = \int_E^{E_0} N_1(E, E') \, R_1(E'|E) \, dE'$$

$$+ \int_0^{E_0-E} M_1(E, E_0 - E') \, R_1(E'|E_0) \, dE' - \beta \frac{\partial N_1(E, E_0)}{\partial E_0}$$

$$(7.8) \quad M_1(E, E_0) \int_0^{E_0} R_2(E'|E_0) \, dE' = \int_0^{E_0-E} R_2(E'|E_0) \, dE'$$

$$+ \int_0^{E} R_2(E'|E_0) \, dE' + \int_0^{E_0} N_1(E, E') \, R_2(E'|E_0) \, dE'$$

$$+ \int_0^{E_0-E} N_1(E, E_0 - E') \, R_2(E'|E_0) \, dE'.$$

To solve (7.7) and (7.8), we use the usual Mellin transform technique, the Mellin transformation being taken with respect to E_0 (*not* E). Defining

$$(7.9) \qquad\qquad N_1(E, s) = \int_E^{\infty} N_1(E, E_0) \, E_0^{s-1} \, dE_0,$$

$$(7.10) \qquad M_1(E, s) = \int_E^\infty M_1(E, E_0) \, E_0^{s-1} \, dE_0,$$

we obtain

$$(7.11) \quad A(1 - s) \, N_1(E, s) = C(1 - s) \, M_1(E, s) + \beta(s - 1) \, N_1(E, s - 1)$$

$$(7.12) \qquad D \, M_1(E, s) = B(1 - s) \, N_1(E, s) + \frac{B(1 - s)}{s} \, E^s.$$

Equation (7.11) can be written as

$$(7.13) \quad A(1 - s) \, N_1(E, s) = C(1 - s) \, M_1(E, s) + \beta(s - 1) \, \delta_s \, N_1(E, s)$$

where δ_s is a unit shift operator operating on the argument s.

Thus the solution of (7.11) and (7.12) can be written to first order in β as

$$(7.14) \quad N_1(E, s) = \frac{1}{D \, A(1 - s) - B(1 - s) \, C \, (1 - s)} \left\{ \frac{C(1 - s) \, B(1 - s)}{s} \, E^s \right.$$

$$+ \left. \frac{D \, \beta \, C(-s) + B(-s) \, E^{s-1}}{D \, A(1 - s) - B(1 - s) \, C(1 - s)} \right\}$$

$$(7.15) \quad M_1(E, s) = \frac{B(1 - s)}{D \, A(1 - s) - B(1 - s) \, C(1 - s)} \left\{ \frac{A(1 - s)}{s} \, E^s \right.$$

$$+ \left. \beta \, \frac{C(-s) \, B(-s)}{D \, A(1 - s) - B(1 - s) \, C(1 - s)} \right\}.$$

The inversion integrals can be evaluated in exactly the same manner as was done in Section 3. We find

$$(7.16) \quad N_1(E, E_0) = (0.443 \, e^y - 1) - 0.1502 \, \frac{D\beta}{E_0} \, y \, e^{2y} \, (1 + 1.358 \, y)$$

$$(7.17) \qquad M_1(E, E_0) = 0.443 \, e^y - \frac{0.116\beta}{E_0} \, e^{2y} \, (y + 1.358 \, y^2).$$

The equations satisfied by $N_2(E, E_0)$ and $M_2(E, E_0)$ can be written down. These can be obtained from (5.24) and (5.25) of Chapter 5 provided we replace $N_2(\mathscr{E})$ by $N_2(E, E_0)$, $N_2(\mathscr{E}/\mathscr{E}')$ by $N_2(E, E')$, $N_2(\mathscr{E}/1 - \mathscr{E}')$ by $N_2(E, E_0 - E')$ and add a term $\beta \, \partial N_2/\partial E_0$ to the right-hand side of (5.24). Thus the calculation of the second factorial moments proceeds on exactly the same lines as in

Section 4 of this chapter. As the method is straightforward we give below the final results for $N_2(E, E_0)$ and $M_2(E, E_0)$ to first order in β:

(7.18) $N_2(E, E_0) = 0.21\, e^{2y} + 0.213\, e^y + 2.253$

$$- \frac{\beta}{E_0}\, [0.069\, e^{4y} - e^{3y}(0.867 - 0.254\, y)$$

$$-e^{2y}(0.89 - 1.005\, y)]$$

(7.19) $M_2(E, E_0) = 0.242\, e^{2y} + 0.213\, e^y$

$$- \frac{\beta}{E_0}\, [0.003\, e^{4y} - e^{3y}(0.459 - 0.132\, y)$$

$$-e^{2y}(0.695 - 0.774\, y)].$$

8. Electromagnetic Cascades in a Polarized Medium

In the cascade theory of electron-photon showers, we have assumed that the mode of development of the electromagnetic cascades is predominantly by bremsstrahlung and pair production. However, the cross-section for bremsstrahlung originally given by Bethe and Heitler [36] is singular with the result that the total cross-section is infinite. This situation known as the infrared catastrophe has been studied extensively in the past, and suitable methods have been devised to deal with the distribution of the number of electrons above a certain energy. However, the cross-section that has been derived by Bethe and Heitler does not take into account the effect of the polarization of the medium. Recently Ter-Mikaelyan (see reference [37]) has studied the effect of the polarization of the medium and obtained the differential cross-section for an electron of energy E to radiate a photon of energy between E' and $E' + dE'$ as

(8.1) $R_1^m(E, E')\, dE' = E'^{-2} \left\{ \frac{4}{3}(E^2 - EE') + E'^2 \right\} \left[1 + \omega^2 \left(\frac{E}{E'} \right)^2 \right]^{-1} \frac{dE'}{E}$

where ω is given by

(8.2) $\omega = \left(\frac{4\pi NZ\, e^2\, \hbar^2}{m^3\, c^4} \right)^{1/2}$

and N is the number of atoms per cm^3, Z being the charge of the nuclei of the medium. The values of ω are 1.9×10^{-4} and 7.5×10^{-5} for air and lead respectively.

We note that the expression for bremsstrahlung given by (8.1) differs from the Bethe-Heitler cross-section by a multiplicative factor that removes the divergence difficulty. The convergent nature of radiation cross-section observed by Harris [38] and the author [39] gives rise to two simplifying features of the problem. The cascade equations, particularly the equations satisfied by the G-functions, can be mathematically justified. The other feature relates to the possibility of explicit solution of the mean numbers at moderately high energies when ionization loss is important.

Timofeev [37] repeated the calculations relating to the mean behavior of showers by using the modified cross-section for bremsstrahlung and estimated the mean number of electrons by the saddle point method. Table 18 gives the

Table 18

			Air				Lead			
s	$A(s)$	$B(s)$	$C(s)$	$\lambda(s)$	$-\lambda'(s)$	$-\mu(s)$	$C(s)$	$\lambda(s)$	$-\lambda'(s)$	$-\mu(s)$
−1.0	−∞	∞	$2.06 \cdot 10^6$	∞	∞	∞	$1.75 \cdot 10^5$	∞	∞	∞
−0.9	−9.73	18.8	$5.24 \cdot 10^5$	3150	32800	3140	7170	372	3600	360
−0.8	−4.63	8.97	$1.37 \cdot 10^5$	1110	10500	1100	3010	166	1160	163
−0.7	−2.86	5.76	$3.68 \cdot 10^4$	461	3840	459	579	87.6	522	85.6
−0.6	−1.94	4.19	$1.02 \cdot 10^4$	207	1690	206	266	49.9	267	48.7
−0.5	−1.36	3.27	$2.93 \cdot 10^3$	98.3	711	97.7	127	28.8	147	29.2
−0.4	−0.0950	2.68	884	48.8	331	48.6	63.5	18.5	84.8	18.4
−0.3	−0.637	2.26	284	25.3	160	26.4	33.5	11.9	63.1	12.1
−0.2	−0.389	1.96	99.3	13.8	80.1	14.2	18.7	7.93	31.3	8.31
−0.1	−0.185	1.73	38.9	7.92	41.4	8.51	11.2	5.42	23.4	6.01
0	0	1.55	17.6	4.84	22.2	5.61	11.2	3.79	12.9	4.57
0.1	0.162	1.40	9.23	3.15	12.6	4.07	7.17	2.72	8.77	3.65
0.2	0.286	1.28	5.58	2.15	7.69	3.21	4.90	1.99	6.12	3.05
0.3	0.407	1.18	3.78	1.53	5.17	2.71	3.55	1.46	4.42	2.64
0.4	0.515	1.10	2.78	1.11	3.55	2.39	2.70	1.08	3.31	2.37
0.5	0.615	1.02	2.17	0.803	2.65	2.19	2.4	0.79	2.55	2.18

modified values of $A(s)$, $C(s)$, $\lambda(s)$, $\mu(s)$, $\lambda'(s)$, and $\mu'(s)$ calculated by Timofeev. Figures 10 and 11 give the mean number of electrons in an electron- and photon-initiated shower in lead and air corresponding to $t = 1$ and 2. The figures also contain the mean numbers estimated on the basis of Bethe-Heitler cross-section. Timofeev has estimated that in the case of lead, the difference in the integral spectra is only 5 per cent of the usual spectrum when $E_0/E = 10^6$. From this, Timofeev has concluded that the polarization of the medium hardly influences the integral spectrum of the shower particles.

The simplifying feature arising from the finiteness of the bremsstrahlung cross-section has been studied by the author [39]. In this case, it is possible

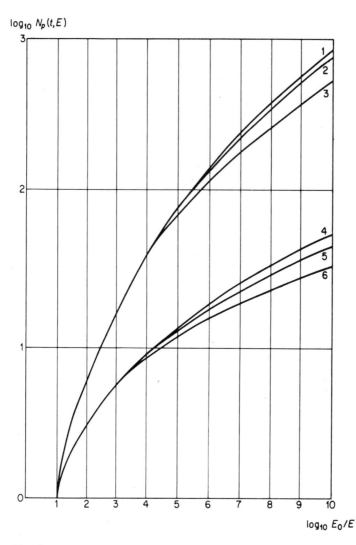

$\log_{10} N_p(t,E)$

$\log_{10} E_0/E$

Fig. 10. Integral spectrum of electrons: 1 — $t = 2$, Bethe-Heitler cross-section; 2 — $t = 2$, Ter-Mikaelyan cross-sections (in air); 3 — $t = 2$, Ter-Mikaelyan cross-sections (in lead); y — $t = 1$, Bethe-Heitler cross-sections; 5 — $t = 1$, Ter-Mikaelyan cross-sections (in air); 6 — $t = 1$, Ter-Mikaelyan cross-sections (in lead).

(Reproduced from *Soviet Physics—JETP*, **14**, (1962) 1064.)

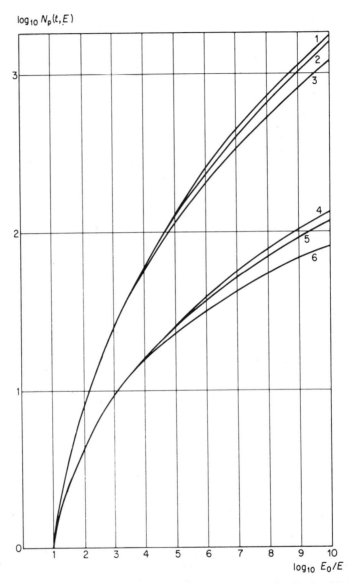

Fig. 11. Integral spectrum of photons: $1 - t = 2$, Bethe-Heitler cross-section; $2 - t = 2$, Ter-Mikaelyan cross-sections (in air); $3 - t = 2$, Ter-Mikaelyan cross-sections (in lead); $4 - t = 1$, Bethe-Heitler cross-section; $5 - t = 1$, Ter-Mikaelyan cross-section (in air); $6 - t = 1$, Ter-Mikaelyan cross-sections (in lead).

(Reproduced from *Soviet Physics—JETP*, **14**, (1962) 1064.)

to introduce the idea of generation and define $F_1^{(n)}(E, E_0, t)$ and $G_1^{(n)}(E, E_0, t)$ as the product densities of degree one of electrons and photons of the n-th generation* that are produced between t and $t + dt$ with energies lying between E and $E + dE$. The usual product densities $F_1(E, E_0, t)$ and $G_1(E, E_0, t)$ are obtained by summing $F_1^{(n)}(E, E_0, t)$ and $G_1^{(n)}(E, E_0, t)$ over all values of n. It is easy to set up integral equations for $F_1^{(n)}(E, E_0, t)$ and $G_1^{(n)}(E, E_0, t)$ and solve them by the transform technique. A noteworthy feature is that the introduction of ionization loss under approximation B does not render the solution difficult.

REFERENCES

1. H. J. Bhabha and W. Heitler, *Proc. Roy. Soc.* (London), **A 159** (1937), 432.
2. J. F. Carlson and R. J. Oppenheimer, *Phys. Rev.*, **51** (1937), 220.
3. D. G. Kendall, *J. Roy. Statist. Soc.*, **B 11** (1949), 230.
4. L. Janossy, *Proc. Phys.. Soc.*, **A 63** (1950), 241.
5. H. J. Bhabha, *Proc. Roy. Soc.* (London), **A 202** (1950), 301.
6. A. Ramakrishnan, *Proc. Camb. Phil. Soc.*, **46** (1950), 595.
7. H. J. Bhabha and A. Ramakrishnan, *Proc. Ind. Acad. Sci.*, **32** (1950), 141.
8. L. Janossy and H. Messel, *Proc. Phys. Soc.* **63 A** (1950), 1101.
9. H. J. Bhabha, *Proc. Ind. Acad. Sci.*, **32** (1950), 154.
10. C. N. Chou and M. Schein, *Phys. Rev.*, **97**, (1955), 206.
11. H. Fay, *Nuovo Cimento*, **5** (1957), 293.
12. W. Heitler, *Quantum Theory of Radiation*, Oxford, 1954.
13. H. J. Bhabha, *Proc. Roy. Soc.* (London), **152 A** (1935), 559.
14. T. Murota, A. Ueda, and H. Tanaka, *Prog. Theor. Phys.*, **16** (1956), 482.
15. A. Ramakrishnan, S. K. Srinivasan, N. R. Ranganathan, and R. Vasudevan, *Proc. Ind. Acad. Sci.*, **45 A** (1957), 311.
16. A. Ramakrishnan and S. K. Srinivasan, *Proc. Ind. Acad. Sci.*, **44 A** (1956), 263.
17. H. J. Bhabha and S. K. Chakrabarti, *Proc. Roy. Soc.* (London), **A 181** (1943), 267.
18. Leonie Janossy and H. Messel, *Proc. Roy. Irish Acad. Sci.*, **54 A** (1951), 217.
19. L. Janossy, *Theory of Cosmic Rays*, Oxford University Press, London, 1950.
20. N. Arley, *Proc. Roy. Soc.* (London), **A 168** (1938), 519.
21. J. C. Butcher, B. A. Chartres, and H. Messel, *Nucl. Phys.*, **6** (1958), 271.
22. S. K. Srinivasan and N. R. Ranganathan, *Proc. Ind. Acad. Sci.*, **45** (1957), 69.
23. A. Ramakrishnan and P. M. Mathews, *Prog. Theor. Phys.*, **11** (1954), 95.
24. S. K. Srinivasan, J. C. Butcher, B. A. Chartres, and H. Messel, *Nuovo Cimento*, **9** (1958), 77.
25. W. T. Scott and G. E. Uhlenbeck, *Phys. Rev.*, **62** (1942), 497.
26. S. K. Srinivasan and K. S. S. Iyer, *Zeit. Phys.*, **182** (1965), 243.

* For an electron-initiated shower $F_1^{(n)} = 0$ for odd n and $G_1^{(n)} = 0$ for even n.

27. E. Fenyves, A. Frenkel, F. Tebiry, J. Perneys, V. Pertryilke, F. Sedlek, and J. Brana, *Nuovo Cimento*, **14** (1959), 1249.
28. S. K. Srinivasan, K. S. S. Iyer, and N. V. Koteswara Rao, *Zeit. Phys.*, **177** (1964), 164.
29. E. Hille, *Analytic Function Theory*, Vol. I, Ginn and Co., New York (1959), p. 254.
30. O. A. Zaimidoroga, Yu. D. Prokoshkin, and V. M. Tsupko-Sitnikov, *Soviet Phys.—JETP* **25** (1967), 51.
31. S. K. Srinivasan and K. S. S. Iyer, *Nuovo Cimento*, **34** (1964), 67.
32. R. F. Christy and S. Kusaka, *Phys. Rev.*, **59** (1941), 414.
33. A. N. Mitra, *Nucl. Phys.*, **3** (1957), 262.
34. H. S. Snyder, *Phys. Rev.*, **76** (1949), 1563.
35. I. P. Ivanenko, *Soviet Phys.—JETP*, **8** (1959), 94.
36. H. A. Bethe and W. Heitler, *Proc. Roy. Soc.* (London), **146** (1934), 83.
37. G. A. Timofeev, *Soviet Phys.—JETP*, **14** (1962), 1062.
38. T. E. Harris, *The Theory of Branching Processes*, Springer-Verlag, Berlin, 1963, p. 169.
39. S. K. Srinivasan, *Nucl. Phys.* **41** (1963), 202.

Chapter 7

EXTENSIVE AIR SHOWERS

1. Historical Remarks

In the previous two chapters, we have dealt with the longitudinal development of electron-photon cascades starting with a single primary electron or photon. As we have mentioned in the introductory remarks of Chapter 5, the primary particles responsible for the generation of electrons and photons are protons that have been accelerated to such fantastic energies probably in the galactic part of the cosmos. These protons by virtue of their extraordinary energy ranging from 10^{13} to 10^{18} eV, make themselves manifest by their high penetrating power. As they reach the upper layers of the atmosphere, they interact with the air nuclei and eject secondary protons, neutrons, and pions. These, in turn, produce secondary reactions interacting with the layers of the atmosphere as a result of which we observe electrons and protons, these particles constituting the "soft component" of a cosmic radiation. Some of the pions (charged ones) decay into muons which by their characteristic inability to interact, penetrate deep into the surface of the earth. The muons, protons, and neutrons constitute the hard component of the cosmic radiation. Such penetrating showers were observed as early as 1938 by Schmeiser and Bothe [1], Auger, Maze, and Grivet-Meyer [2], Janossy and Lowell [3]. In the year 1949, Heitler and Janossy [4] motivated by the work of Brown *et al.* [5], estimated the size-frequency distribution of penetrating showers.* Later on, Messel and Potts [8] attempted an explicit solution for the integral spectra of various types of particles. This was followed by Ueda and Ogita [9] and Olbert and Stora [10] who have investigated the structure of extensive air showers on the basis of certain plausibly correct assumptions of nucleon-nucleon collision cross-sections. An analysis of the numerical results and a

* As we have observed in the introductory remarks of Chapter 5, innumerable accounts of experimental data and perhaps plausible theoretical investigations were published during the late forties and early fifties and it is not possible to give an adequate account of the historical evolution in a chapter of this kind. Thus we have confined ourselves to those papers which are fundamentally important from the viewpoint of stochastic processes. The interested reader is referred to the critical review articles by Messel [6] and Greisen [7].

comparison of the experimental data has been carried out by Fukuda, Ogita, and Ueda [11], Ueda and McCusker [12], and Ueda [13]. From these investigations, the mean behavior of these showers is fairly well established at least on the basis of a phenomenological cross-section for nucleon-nucleon collisions. There are one or two pertinent results that deal with the fluctuation in the size due to the fluctuation in position of the first interaction of the primary or some of the secondaries. The object of this chapter is to present the relevant stochastic features of these problems and point out some of the difficult aspects. We shall not deal with the angular and lateral spreads of such showers, although they are quite important, and considerable work has been done in this direction especially during the past few years. The motivation to do so has stemmed from the situation that although mean angular and lateral spectra of particles have been calculated, much remains to be done in the direction of determining the number distribution.

2. Nucleon Component

As we have mentioned earlier, the primary cosmic radiation consists predominantly of protons and so the development of the nucleon component constitutes the first phase of an extensive air shower. In this section, we shall deal with the distribution of the number of nucleons generated by a single proton of energy E_0. We shall assume the probability that a nucleon of energy E gives rise to secondary and recoil nucleons of energies in the interval $(E_1, E_1 + dE_1)$ and $(E_2, E_2 + dE_2)$ is

$$(2.1) \quad \phi(E_0; E_1, E_2) \, dE_1 \, dE_2 = \phi(\mathscr{E}_1, \mathscr{E}_2) \, d\mathscr{E}_1 \, d\mathscr{E}_2$$
$$(\mathscr{E}_2 = E_1/E_0, \; \mathscr{E}_2 = E_2/E_0)$$

with $E_0 = E_1 + E_2 + E_m$, E_m being the energy that goes into the production of mesons. It follows from (2.1) that

$$(2.2) \quad \int_0^1 \int_0^{1-\mathscr{E}_1} \phi(\mathscr{E}_1, \mathscr{E}_2) \, d\mathscr{E}_1 \, d\mathscr{E}_2 = \text{a constant} = k.$$

A number of general properties of the function ϕ can be deduced once we assume the cross-section to be of the form (2.1):

(i) ϕ must be symmetrical in the variables $\mathscr{E}_1$, $\mathscr{E}_2$ since the recoil and secondary nucleons cannot be distinguished

(ii) ϕ must vanish whenever $\mathscr{E}_1$ or $\mathscr{E}_2$ tends to zero or unity since we postulate complete inelasticity.

On the basis of these two assumptions, Messel [14] assumed ϕ to be given by

$$(2.3) \qquad \phi(\mathscr{E}_1, \mathscr{E}_2) = \sigma(\mathscr{E}_1 \mathscr{E}_2)^{\beta} (1 - \mathscr{E}_1 - \mathscr{E}_2)^{\delta}$$

where β and δ are certain constants to be determined on the basis of experimental data on the nucleon component of cosmic radiation.

There are a number of details which we do not propose to deal with. For example, we have assumed that the cross-section for nucleon-nucleon collision is the same whether the ejected particle is a neutron or proton. This is not a correct and justifiable assumption since the hypothesis of charge independence of nuclear forces which dominates the world of elementary particles will give definite relationships between the cross-section for collision between the two types of nucleons. Thus in a realistic and complete treatment, it may be necessary to distinguish between protons and neutrons. We next wish to observe that in a high energy encounter, strange particles may be produced and these, in turn, may interact producing a cascade or may spontaneously decay through the medium of weak interaction. To propose a comprehensive theory, we must take into account a number of other phenomena of high complexity falling into the realms of fundamental particle physics and it is precisely because of the very complexity of the problem that we have chosen a simple model to describe the general features of an air shower.

2.1. PRODUCT DENSITIES

Let $f_1(\mathscr{E}; t)$ and $f_2(\mathscr{E}_1, \mathscr{E}_2; t)$ be the product densities of degrees one and two of nucleons that exist at t. To obtain differential equations satisfied by f_1 and f_2, we increase t by Δ and investigate the possibilities in the interval $(t, t + \Delta)$. Taking into account the collisions we find

$$(2.4) \qquad \frac{\partial f_1(\mathscr{E}; t)}{\partial t} = -k f_1(\mathscr{E}; t) + 2 \int \int f_1\left(\frac{\mathscr{E}}{\mathscr{E}'}; t\right) \phi(\mathscr{E}', \mathscr{E}'') \frac{1}{\mathscr{E}'} d\mathscr{E}' \, d\mathscr{E}''$$

$$(2.5) \qquad \frac{\partial f_2(\mathscr{E}_1, \mathscr{E}_2; t)}{\partial t} = - 2k f_2(\mathscr{E}_1, \mathscr{E}_2; t) + 2 \int \int \left[f_2\left(\mathscr{E}, \frac{\mathscr{E}_2}{\mathscr{E}'}; t\right) \right.$$
$$\left. + f_2\left(\frac{\mathscr{E}_1}{\mathscr{E}'}, \mathscr{E}_2; t\right) \right] \phi(\mathscr{E}', \mathscr{E}'') \frac{d\mathscr{E}' \, d\mathscr{E}''}{\mathscr{E}'} + 2 \int f_1(\mathscr{E}, t) \phi\left(\frac{\mathscr{E}_1}{\mathscr{E}}, \frac{\mathscr{E}_2}{\mathscr{E}}\right) \frac{d\mathscr{E}}{\mathscr{E}^2}.$$

We have assumed that nucleons are produced only by nucleons. However, the pions that are ejected in a nucleon-nucleus collision in turn produce nucleons as a result of their collisions with the nucleons. If we add an extra term on the right-hand side of (2.4) and (2.5) due to such collisions, then we obtain coupled equations. We shall discuss this in the next section when we deal with the pions and muons.

Defining the Mellin transforms of $f_1(\mathscr{E}; t)$ and $f_2(\mathscr{E}_1, \mathscr{E}_2; t)$ as

$$(2.6) \qquad c_1(s; t) = \int_0^1 f_1(\mathscr{E}; t)\, \mathscr{E}^{s-1}\, d\mathscr{E}$$

$$(2.7) \qquad c_2(s_1, s_2; t) = \int_0^1 \int_0^1 f_2(\mathscr{E}_1, \mathscr{E}_2; t)\, \mathscr{E}_1^{s_1-1}\, \mathscr{E}_2^{s_2-1}\, d\mathscr{E}_1\, d\mathscr{E}_2$$

we obtain

$$(2.8) \qquad \frac{\partial\, c_1(s; t)}{\partial t} = -c_1(s, t)\,(k - a(s))$$

$$(2.9) \qquad \frac{\partial\, c_2(s_1, s_2; t)}{\partial t} = -c_2(s_1, s_2; t)\,[2k - a(s_1) - a(s_2)]$$

$$+ c_1(s_1 + s_2 - 1; t)\, b(s_1, s_2)$$

where

$$(2.10) \qquad b(s_1, s_2) = 2 \int_0^1 \int_0^{1-\mathscr{E}_1} \phi(\mathscr{E}_1, \mathscr{E}_2)\, \mathscr{E}_1^{s_1-1}\, \mathscr{E}_2^{s_2-1}\, d\mathscr{E}_1\, d\mathscr{E}_2$$

$$a(s_1) = b(s_1, 1).$$

We shall assume that at $t = 0$, there is a single nucleon of energy E_0 so that

$$(2.11) \qquad f_1(\mathscr{E}; 0) = \delta(\mathscr{E} - 1), \qquad f_2(\mathscr{E}_1, \mathscr{E}_2; 0) = 0.$$

Solving equations (2.8) and (2.9), we obtain

$$(2.12) \qquad c_1(s, t) = e^{-A(s)t}$$

$$c_2(s_1, s_2; t) = b(s_1, s_2)\, \frac{[e^{-A(s_1+s_2-1)t} - e^{-[A(s_1)+A(s_2)]t}]}{A(s_1) + A(s_2) - A(s_1 + s_2 - 1)}$$

where $A(s) = k - a(s)$.

Thus the mean and mean square number of nucleons above a fraction are given by

$$(2.13) \qquad E\{N(\mathscr{E}_c; t)\} = \frac{1}{2\pi i} \int_{\sigma-i\infty}^{\sigma+i\infty} \frac{c_1(s; t)\, e^{y(s-1)}\, ds}{s - 1}$$

$$(2.14) \quad E\{[N(\mathscr{E}_c; t)]^2\} = \frac{1}{2\pi i} \int\limits_{\sigma-i\infty}^{\sigma+i\infty} \frac{c_1(s; t)\, e^{y(s-1)}}{s-1}\, ds$$

$$+ \left(\frac{1}{2\pi i}\right)^2 \int\limits_{\sigma+i\infty}^{\sigma+i\infty} \int\limits_{\sigma-i\infty}^{\sigma-i\infty} \frac{c_2(s_1, s_2; t)\, e^{y(s_1+s_2-2)}}{(s_1-1)(s_2-1)}\, ds_1\, ds_2$$

where $y = -\log \mathscr{E}_c$.

2.2. TWO SIMPLE MODELS

The mean and mean square deviation of the number of nucleons produced in a cascade generated by a single primary have been explicitly calculated by Ramakrishnan and Srinivasan [15] for two special forms of $\phi(\mathscr{E}_1, \mathscr{E}_2)$. They have neglected the formation of pions so that ϕ is of the form

$$(2.15) \qquad \phi(\mathscr{E}_1, \mathscr{E}_2) = \phi(\mathscr{E}_1, 1 - \mathscr{E}_1)\, \delta(\mathscr{E}_2 - \overline{1 - \mathscr{E}_1})$$
$$= \omega(\mathscr{E}_1)\, \delta(\mathscr{E}_2 - \overline{1 - \mathscr{E}_1}).$$

In the first model, $\omega(\mathscr{E}) = k = $ constant. In this case, we have

$$(2.16) \qquad\qquad a(s) = \frac{2}{s}.$$

It can be shown that $E\{N(\mathscr{E}_c; t)\}$ has a maximum for a given y at $t = t_y = 2(y-1)/k$, the maximum value being given by

$$(2.17) \qquad\qquad E\{N(\mathscr{E}_c; t)\}_{max} = e^y/\sqrt{2\pi y}.$$

The numerical value of the mean and mean square number of nucleons for various t have been obtained by Ramakrishnan and Srinivasan by the use of saddle point formula. Tables 19 and 20 give the value of $E\{N(y; t)\}$ and $E\{[N(y; t)]^2\}$. For convenience, we have also plotted the ratio of the mean square deviation to the mean square deviation (Table 21) corresponding to a Poisson and Furry process (discussed in Chapter 1).

In the second model, Ramakrishnan and Srinivasan have assumed ω to be given by

$$(2.18) \qquad\qquad \omega(\mathscr{E}) = \frac{1}{4}\left(\frac{1}{\mathscr{E}} + \frac{1}{1-\mathscr{E}}\right).$$

The total cross-section in this case is no longer finite. The model has been chosen in this manner as an analogy with the radiation cross-section of an electron. Of course, in the case of a nucleon-nucleon collision, the cross-section is always finite in view of the non-zero mass of the pions mediating the nucleon-nucleon interaction. The numerical results are of some importance since this will bring out one important property of the multiplicative

Table 19

$$E\{N(y;t)\}$$

$[E\{N(y;t)\}]^2$ is given in parentheses

y	t						
	1	2	3	4	5	6	8
2	1.74 (3.02)	2.08 (4.34)	1.98 (3.90)	1.94 (3.75)			
3	2.11 (4.4b)	3.24 (10.5)	4.14 (17.2)	4.63 (21.4)	4.67 (21.9)	4.27 (18.2)	3.21 (10.3)
4		4.29 (18.4)	6.51 (42.4)	8.56 (73.3)	10.1 (101)	10.9 (119)	10.4 (108)
5		5.13 (26.3)	8.94 (79.9)	13.5 (181)	18.1 (327)	22.7 (515)	26.5 (701)
6			11.3 (127)	18.9 (358)	28.4 (808)	38.7 (1500)	56.8 (3230)
7				24.5 (600)	40.5 (1640)	60.7 (3690)	107 (11500)
8					53.7 (2880)	87.6 (7670)	182 (33100)
9						120 (14300)	283 (79800)
10							415 (173×10^3)
11							573 (328×10^3)

processes namely that the spread of the showers about their mean size does not depend on the finite nature of the cross-section. Tables 22, 23, and 24 contain the mean, mean square, and mean square deviation of the number of nucleons as well as the ratio of the mean square deviation to the corresponding Poisson and Furry values.

Table 20
$E\{N(y;t)\}^2$
$\sigma^2(y;t)$ is given in parentheses

y	t						
	1	2	3	4	5	6	8
2	3.92	4.38	7.78	6.63			
	(0.90)	(0.04)	(3.88)	(2.88)			
3	5.83	13.3	20.2	24.3	24.3	21.2	13.6
	(1.38)	(2.8)	(3.1)	(2.9)	(2.5)	(3.0)	(3.3)
4		25.4	54.1	87.3	115	129	116
		(7.0)	(11.7)	(14.0)	(14)	(11)	(7.2)
5		40.4	110	232	390	577	741
		(14.1)	(30)	(51)	(63)	(62)	(40)
6			184	486	1080	1770	3550
			(57)	(130)	(280)	(280)	(320)
7				860	2210	4630	12900
				(260)	(570)	(940)	(1500)
8				4040	102×10^2	398×10^2	
				(1160)	(25×10^2)	(67×10^2)	
9					194×10^2	101×10^3	
					(51×10^2)	(21×10^3)	
10						225×10^3	
						(52×10^3)	
11						446×10^3	
						(118×10^3)	

Table 21

σ^2/σ_F^2

σ^2/σ_P^2 **is given in parentheses**

y	t						
	1	2	3	4	5	6	8
2	0.71	0.028	2.0	1.6			0.37
	(0.52)	(0.017)	(2.0)	(1.5)			(0.88)
3	0.59	0.44	0.23	0.17	0.14	0.22	0.073
	(0.65)	(0.85)	(0.74)	(0.63)	(0.51)	(0.71)	(0.69)
4		0.47	0.33	0.22	0.16	0.10	0.059
		(1.6)	(1.8)	(1.6)	(1.5)	(0.99)	(1.5)
5		0.67	0.42	0.30	0.20	0.13	0.10
		(2.8)	(3.3)	(3.7)	(2.9)	(2.7)	(5.7)
6			0.49	0.38	0.28	0.19	0.13
			(5.1)	(6.8)	(7.7)	(7.1)	(14)
7				0.45	0.35	0.26	0.20
				(11)	(14)	(16)	(37)
8					0.41	0.33	0.27
					(22)	(28)	(75)
9						0.36	0.30
						(42)	(130)
10							0.36
							(200)
11							

Table 22
$E\{N(y;t)\}$
$[E\{N(y;t)\}]^2$ is given in parentheses

y	t						
	2	3	4	5	6	7	8
2	1.96 (3.83)	1.66 (2.76)	1.26 (1.59)				
3	3.79 (14.3)	4.45 (19.8)	3.93 (15.5)	3.26 (10.6)	2.45 (6.00)	1.70 (2.89)	
4	6.51 (42.4)	8.63 (74.5)	9.59 (92.0)	9.38 (88.0)	8.35 (69.7)	6.74 (45.4)	5.11 (26.1)
5	10.4 (108)	16.0 (256)	20.5 (420)	22.8 (520)	22.9 (524)	21.1 (445)	18.1 (328)
6		27.8 (773)	39.9 (1590)	49.9 (2490)	55.7 (3100)	56.9 (3230)	53.9 (2910)
7		45.5 (2070)	72.8 (5300)	100 (10^4)	123 (151×10^2)	138 (190×10^2)	142 (202×10^2)
8			128 (164×10^2)	191 (365×10^2)	254 (645×10^2)	308 (949×10^2)	345 (119×10^3)
9			214 (458×10^2)	346 (120×10^3)	499 (249×10^3)	647 (419×10^3)	778 (605×10^3)
10			348 (121×10^3)	605 (366×10^3)	931 (867×10^3)	1280 (164×10^4)	1660 (276×10^4)
12					2950 (868×10^4)	4540 (206×10^5)	6620 (438×10^5)
14						144×10^2 (207×10^6)	231×10^2 (534×10^6)

Table 23
$E\{N(y;t)\}^2$
$\sigma^2(y;t)$ is given in parentheses

y	\multicolumn{7}{c}{t}						
	2	3	4	5	6	7	8
2	4.41 (0.58)	3.6 (0.8)	2.24 (0.65)				
3	16.9 (2.6)	20.0 (0.2)	17.3 (1.8)	12.4 (1.8)	7.77 (1.77)	4.50 (1.11)	
4	51.4 (9.0)	83.9 (9.4)	99.8 (7.8)	94.0 (6.0)	78.0 (8.3)	51.1 (5.6)	31.8 (5.7)
5	132 (24)	293 (37)	459 (39)	552 (32)	546 (22)	464 (19)	345 (17)
6		890 (117)	1760 (170)	2660 (170)	3200 (100)	3300 (70)	2960 (50)
7		2470 (400)	6000 (700)	109×10^2 (9×10^2)	159×10^2 (8×10^2)	196×10^2 (6×10^2)	205×10^2 (3×10^2)
8			185×10^2 (21×10^2)	403×10^2 (38×10^2)	685×10^2 (40×10^2)	992×10^2 (43×10^2)	121×10^3 (3×10^3)
9			516×10^2 (58×10^2)	133×10^3 (12×10^3)	265×10^3 (16×10^3)	446×10^3 (27×10^3)	631×10^3 (26×10^3)
10			139×10^3 (18×10^3)	417×10^3 (51×10^3)	939×10^3 (72×10^3)	180×10^4 (16×10^4)	290×10^4 (14×10^4)
12					966×10^4 (98×10^4)	233×10^5 (27×10^5)	473×10^5 (35×10^5)
14						234×10^6 (27×10^6)	585×10^6 (51×10^6)

Table 24
$$\sigma^2/\sigma_P^2$$
σ^2/σ_P^2 is given in parentheses

y	t = 2	3	4	5	6	7	8
2	0.31 (0.30)	0.72) (0.48)	2.0 (0.52)				
3	0.25 (0.69)	0.05 (0.05)	0.16 (0.46)	0.25 (0.55)	0.50 (0.72)	0.93 (0.65)	
4	0.24 (1.4)	0.14 (1.1)	0.095 (0.81)	0.076 (0.64)	0.14 (0.99)	0.14 (0.83)	0.27 (1.1)
5	0.25 (2.3)	0.15 (2.3)	0.10 (1.9)	0.064 (1.4)	0.044 (0.96)	0.044 (0.90)	0.055 (0.94)
6		0.16 (4.2)	0.11 (4.3)	0.070 (3.4)	0.032 (1.8)	0.022 (1.2)	0.017 (0.94)
7		0.20 (8.8)	0.13 (9.6)	0.091 (9.0)	0.053 (6.5)	0.032 (4.3)	0.015 (2.1)
8			0.13 (16)	0.10 (20)	0.062 (16)	0.045 (14)	0.025 (8.7)
9			0.13 (27)	0.11 (38)	0.065 (32)	0.065 (42)	0.043 (33)
10			0.15 (52)	0.14 (84)	0.083 (77)	0.10 (130)	0.051 (84)
12					0.11 (330)	0.13 (590)	0.079 (530)
14						0.13 (19×10^2)	0.096 (22×10^2)

2.3. NUCLEON CASCADES GENERATED BY A SPECTRUM OF PARTICLES

The results presented in Sections 2.1 and 2.2 are valid for a shower generated by a single primary. However, experimentally, it is the mean spectrum that can be measured. For example, if we use the primary power law spectrum as given in reference [6], then the initial condition (2.11) should be modified as

$$(2.19) \qquad \begin{aligned} f_1(E, 0) &= \gamma E_c^\gamma/E^{\gamma+1} & E > E_c \\ &= 0 & E < E_c \end{aligned}$$

where E_c is the latitude cut-off energy of primary protons so that

$$(2.20) \qquad c_1(s; 0) = \frac{E_c^{s-1}}{\gamma + 1 - s}.$$

The complete Mellin transform solution is given by

(2.21)
$$c_1(s; t) = \frac{\gamma E_c^{s-1}}{\gamma + 1 - s} e^{-A(s)t}.$$

Then $f_1(E, t)$ can be readily obtained for $E \geqslant E_c$. For, in this case, we obtain

(2.22)
$$f_1(E, t) = \frac{1}{2\pi i} \int_{\sigma - i\infty}^{\sigma + i\infty} \frac{\gamma}{\gamma + 1 - s} \left(\frac{E_c}{E}\right)^{s-1} e^{-A(s)t} ds$$

$$= \frac{\gamma}{E} \left(\frac{E_c}{E}\right)^{\gamma} e^{-A(\gamma+1)t}.$$

However, when $E < E_c$, the inversion is not obvious for then we have to take into account the residue arising out of the singularities due to the term $e^{-A(s)t}$. It is to be noted that when the mean spectrum is specified, the product density of degree two is not determined since this will depend on the initial condition for f_2. In this case, we cannot assert that $f_2(E_1, E_2; 0) = 0$ as in the case of a shower excited by a single primary. *This point seems to have escaped the attention of both theorists and experimentalists.* We shall have occasion to deal with this point again in Section 4.

3. Mean Behavior of Extensive Air Showers

The mean behavior of the different components of an extensive air shower has been investigated in the first instance by Messel and Potts [8] and later by Ueda *et al.* [9, 11, 12, 13]. We shall deal with some of the salient features of these calculations.

Let $q_i(E, E_0, x)$ be the product density of degree one of particles of type i at a depth x gm|cm^2 in a shower generated by a single primary proton of energy E_0. Let $i = 1, 2, 3, 4, 5, 6$ stand respectively for nucleons, charged pions, neutral pions, muons, electrons, and photons. It is assumed that

(i) nucleons and pions are produced only in nucleon-nucleon collisions;
(ii) muons are produced by the decay of charged pions;
(iii) photons are produced in pairs by the decay of neutral pions and singly by radiation of electrons through bremsstrahlung, and
(iv) electrons are produced in pairs by photons.

The cross-sections are defined as follows:

(i) $W_1(E_1, E) dE$: the differential cross-section for the production of a nucleon of energy between E and $E + dE$ by an incident nucleon of energy E_1.

(ii) $W_2(E_1, E)\, dE$: differential cross-section for the production of charged pion of energy between E and $E + dE$ by an incident nucleon of energy E_1. (For a neutral pion, we multiply this by the factor $1/a$.)

(iii) $W_3(E_1, E)$ and $W_4(E_1, E)$: the usual differential cross-section for bremsstrahlung and pair production.

The mean lifetimes of neutral pions and muons are assumed to be $\tau(\pi^\circ)$ and $\tau(\mu)$ seconds in the center of mass frame of reference. Let $l(n)$ and $l(\pi)$ denote the interaction mean free paths of nucleon and charged pions in gm/cm². We shall use a notation $l(n, \pi) = l(n)/l(\pi)$, $l(\pi, n) = l(\pi)/l(n)$. Finally, let k be the radiation unit of electrons and photons measured in gm/cm², c the velocity of light, and $m(\pi)$ $m(\pi^\circ)$ and $m(\mu)$ the masses of charged pions, neutral pions, and muons respectively.

3.1. DIFFUSION EQUATIONS

The diffusion equations, under the assumptions stated above, are given by

$$(3.1) \quad \left\{ l(n) \frac{\partial}{\partial x} + 1 \right\} q_1(E, E_0, x) = \int_E^\infty q_1(E_1, E_0, x)\, W_1(E_1, E)\, dE_1,$$

$$(3.2) \quad \left\{ l(\pi) \frac{\partial}{\partial x} - \frac{m(\pi)\, cl(\pi)}{\tau(\pi) Ex'(t)} - l(\pi)\, \beta_\pi \frac{\partial}{\partial E} + 1 \right\} q_2(E, E_0, x)$$
$$= 1(\pi, n) \int_E^\infty q_2(E_1, E_0, x)\, W_2(E_1, E)\, dE_1,$$

$$(3.3) \quad \left\{ 1(\pi) \frac{\partial}{\partial x} - \frac{m(\pi^\circ)\, cl(\pi)}{\tau(\pi^\circ)\, Ex'(t)} \right\} q_3(E, E_0, x) = \frac{l(\pi, n)}{a}$$
$$\int_E^\infty q_1(E_1, E_0, x)\, W_2(E_1, E)\, dE_1,$$

$$(3.4) \quad \left\{ \frac{\partial}{\partial x} - \frac{m(\mu)c}{\tau(\mu) Ex'(t)} - \beta_\mu \frac{\partial}{\partial E} \right\} q_4(E, E_0, x) = \int_E^\infty q_3(E_1, E_0, x)$$
$$\left\{ \frac{m(\pi)^2}{m(\pi)^2 - m(\mu)^2} \right\} \left\{ \frac{-m(\pi)c}{\tau(\pi)\, E_1 x'(t)} \right\} \frac{dE_1}{E_1},$$

$$(3.5) \quad \left\{ k \frac{\partial}{\partial x} + \int_0^E W_3(E, E')\, dE' \right\} q_5(E, E_0, x)$$

$$= \int_E^\infty q_5(E_1, E_0, x)\, W_4(E_1 - E, E)\, dE_1$$

$$+ \int_E^\infty q_5(E_1, E_0, x)\, W_3(E, E_1 - E)\, dE_1,$$

$$(3.6) \quad \left\{ k \frac{\partial}{\partial x} + \int_0^E W_4(E, E')\, dE' \right\} q_6(E, E_0, x)$$

$$= \int_E^\infty q_5(E_1, E_0, x)\, W_3(E_1, -E, E)\, dE_1$$

$$+ 2k \int_E^\infty q_3(E_1, E_0, x) \left\{ \frac{-m(\pi^\circ)\, c}{\tau(\pi^\circ)\, E_1\, x'(t)} \right\} \frac{dE_1}{E_1}$$

where β_π and β_μ are the energy losses due to ionization in MeV per gram per square centimeter, by charged pions and muons. We have neglected the ionization loss suffered by electrons and this is a valid assumption if we confine our attention to high energy electrons. We have also neglected the contribution of electrons from high energy muons since this is very small.

3.2. NUCLEON SPECTRUM

For a primary spectrum of protons of the form (2.19), the mean numbers of nucleons at any depth can be calculated. If $n_1(E, E_c, x)$ is the mean number of nucleons above an energy $E > E_c$, then

$$n_1(E, E_c, x) = \left(\frac{E_c}{E} \right)^\gamma \exp\left[-h(\gamma) x / l(n) \right]$$

where

$$h(s) = 1 - \iint \left(\frac{E}{E_1} \right)^s W_1(E_1, E)\, dE\, dE_1.$$

Thus for a primary integral spectrum $(180/E_0)^{1.5}$, we have

$$n_1(E, E_c, x) = \left(\frac{180}{E} \right)^{1.5} e^{-x}$$

where x is measured in units of $l(n)/n(\gamma) \simeq 100$ gm/cm^2.

3.3 PION AND MUON SPECTRUM

The pion and muon spectrum can be obtained by solving equations (3.2), (3.3), and (3.4). This has been done by Messel and Potts for a case when the initial spectrum of protons is given by (2.19). They have neglected the ionization loss ($\beta_\pi = \beta_\mu = 0$) and tabulated the results for various energies and depths. Table 25 and 26 give the mean number of charged pions and muons.

Table 25. Average Number of Charged Pions with Energy $> E$ GeV at depth x (in 100 gm/cm²) in the Atmosphere, Assuming a Nucleon Primary Integral Spectrum $(180/E_0)^{1\cdot5}$

E					x				
	0	0.5	1	1.5	2	3	4	6	10
10	0	0.703	0.928	0.927	0.832	0.601	0.399	0.197	0.0914
15	0	0.534	0.716	0.728	0.666	0.492	0.347	0.185	0.0908
20	0	0.434	0.590	0.608	0.565	0.430	0.313	0.176	0.0903
30	0	0.319	0.442	0.465	0.442	0.353	0.269	0.164	0.0893
60	0	0.177	0.254	0.278	0.274	0.238	0.197	0.137	0.0847

Table 26. Average Number of Muons with Energy $> E$ GeV at Depth x (in 100 gm/cm²) in the Atmosphere, Assuming a Nucleon Primary Integral Spectrum $(180/E_0)^{1\cdot5}$

E					x				
	0	0.5	1	1.5	2	3	4	6	10
10	0	6.44	10.6	13.3	15.1	17.0	17.9	18.5	18.7
15	0	3.31	5.48	6.92	7.93	8.98	9.50	9.91	10.1
20	0	2.03	3.40	4.31	4.94	5.67	6.03	6.34	6.49
30	0	1.01	1.70	2.18	2.53	2.92	3.13	3.33	3.46
60	0	0.291	0.497	0.645	0.759	0.895	0.978	1.07	1.14

Table 27.   Average Number of Electrons with Energy $> E$ GeV at Depth x (in 100 gm/cm²) in the Atmosphere, Assuming a Nucleon Primary Integral Spectrum $(180/E_0)^{1.5}$

E	x								
	0	0.5	1	1.5	2	3	4	6	10
10	0	1.04	1.60	1.66	1.48	0.956	0.540	0.144	0.00771
15	0	0.566	0.870	0.905	0.806	0.521	0.294	0.7083	0.00420
20	0	0.368	0.565	0.568	0.524	0.338	0.191	0.0508	0.00273
30	0	0.200	0.308	0.320	0.285	0.184	0.104	0.0277	0.00148
60	0	0.0707	0.109	0.113	0.101	0.0651	0.0367	0.00979	0.00525

3.4.   ELECTRONS AND PHOTONS

Equations (3.5) and (3.6) can be solved by the usual methods. In fact, they are the usual equations of electromagnetic cascades except for the fact that there is a contribution to the photon differential spectrum arising from the neutral pion decay. Since the method is straightforward, we have not gone

Table 28.   Average Number of Photons with Energy $> E$ GeV at Depth x (in 100 gm/cm²) in the Atmosphere, Assuming a Nucleon Primary Integral Spectrum $(180/E_0)^{1.5}$

E	x								
	0	0.5	1	1.5	2	3	4	6	10
10	0	2.69	3.27	3.06	2.59	1.58	0.869	0.226	0.0119
15	0	1.46	1.78	1.67	1.41	0.860	0.473	0.123	0.00648
20	0	0.951	1.16	1.08	0.914	0.559	0.307	0.0798	0.00421
30	0	0.517	0.629	0.590	0.498	0.304	0.167	0.0435	0.00229
60	0	0.183	0.222	0.208	0.176	0.108	0.0591	0.0154	0.000810

into the details of the solution. Tables 27 and 28 give the mean number of electrons and photons above E_c. Table 29 gives the ratio of average number of various components of the extensive air shower.

Table 29. Ratios of Average Numbers of Various Components for Different Values of the Depth x (in 100 gm/cm²) in the Atmosphere and for $E > 10$ GeV

Ratio	x								
	0	0.5	1	1.5	2	2	4	6	10
$(\pi + \mu)/T$	0	0.15	0.40	0.82	1.6	4.5	0.13	96	5.300
e/p	—	0.39	0.49	0.54	0.57	0.60	0.62	0.64	0.65
e/T	0	0.022	0.057	0.098	0.15	0.25	0.39	0.76	2.2
p/T	0	0.058	0.12	0.18	0.26	0.42	0.62	1.2	3.4
$e/(T + \pi + \mu)$	0	0.019	0.039	0.053	0.057	0.045	0.027	0.0076	0.00041
$p/(T + \pi + \mu)$	0	0.051	0.081	0.098	0.10	0.074	0.044	0.012	0.00063

3.5. DISCUSSION OF THE MODEL

The results presented so far have been obtained under certain very special circumstances the validity of which we wish to discuss. We have neglected the production of nucleons by energetic secondary pions. This is indeed a serious drawback of the model. There is yet another feature that has not been taken into account. In an energetic nucleon-nucleus collision, normally more than one nucleon and pion will be emitted. Thus an increase in the multiplicity of production is likely to change the character of the mean size of the extensive air shower. Ueda and Ogita [9] have taken into account these features. The main assumptions of their model of nucleon-nucleon and pion-nucleon collisions are as follows:

(i) The collision cross-section is independent of the energy of a colliding particle and also independent of the type of colliding particle (i.e. nucleon or charged pion).

(ii) The number of secondary particles produced by a particle of energy E in the laboratory system is given by

$$n = \alpha(\eta E)^\gamma$$

where α and γ are suitable parameters and η represents the inelasticity so that an energy of amount ηE is transferred to secondary particles and $(1 - \eta)E$ is carried away by the colliding particle.

(iii) The energy transferred to secondary particles is equi-partitioned among them.

(iv) Two of the secondary particles are nucleons. Of the remaining $(n - 2)$ particles, $\Delta(n - 2)$ are charged pions and $(1 - \Delta)(n - 2)$ particles are neutral pions. The fraction Δ is a constant. The production of nucleon anti-nucleon pairs and heavy mesons are neglected.

(v) Nuclear active particles of energy lower than the critical value E_c, do not contribute to the further development of the nucleonic cascade process in the atmosphere.

Table 30. The Numbers of Nucleonic Components at Mountain Altitude (720 gm cm^{-2}) and Sea Level (1033 gm cm^{-2})*

	E_0					
Depth	720 gm cm^{-2}			1033 gm cm^{-2}		
η	10^4	10^6	S	10^4	10^6	S
1	1.2	$6.5 \cdot 10^2$	1.4	$9.0 \cdot 10^{-2}$	$7.9 \cdot 10$	1.5
0.5	8.4	$1.4 \cdot 10^3$	1.1	1.7	$4.5 \cdot 10^2$	1.2
0.3	$2.0 \cdot 10$	$2.3 \cdot 10^3$	1.0	$1.4 \cdot 10$	$1.8 \cdot 10^3$	1.1

* S is defined by the relation (between the primary energy E_0 and N the number of nuclear active particles) $N \simeq E_0^S$. Attenuation lengths are obtained from the numbers at both altitudes.

Table 31. The numbers of Electrons Having Energies Larger than 10 Mc^2 at Both Altitudes

	E_0			
Depth	720 gm cm^{-2}		1033 gm cm^{-2}	
η	10^4	10^6	10^4	10^6
1	1.2 $(3.6 \cdot 10^{-4})$*	$9.1 \cdot 10^{+2}$ $(9.5 \cdot 10^{-4})$	$4.0 \cdot 10^{-2}$ $(1.4 \cdot 10^{-4})$	$5.3 \cdot 10^2$ $(3.8 \cdot 10^{-4})$
0.5	2.2 $(6.5 \cdot 10^{-4})$	$1.2 \cdot 10^3$ $(1.4 \cdot 10^{-3})$	$2.0 \cdot 10^{-1}$ $(3.6 \cdot 10^{-4})$	$1.7 \cdot 10^2$ $(6.8 \cdot 10^{-4})$
0.3	4.6 $(1.2 \cdot 10^{-3})$	$1.7 \cdot 10^3$ $(2.3 \cdot 10^{-3})$	1.1 $(9.2 \cdot 10^{-4})$	$5.2 \cdot 10^2$ $(1.3 \cdot 10^{-3})$

* The values in the parentheses are ratios of the numbers of such electrons to the numbers of total electrons.

Table 32. The Number of Muons at Both Altitudes

| Depth | E_0 | | | |
| | 720 gm cm^{-2} | | 1033 gm cm^{-2} | |
η	10^4	10^6	10^4	10^6
1	$1.3 \cdot 10^2$	$8.4 \cdot 10^3$	$1.4 \cdot 10^2$	$8.7 \cdot 10^3$
0.5	$9.6 \cdot 10$	$6.2 \cdot 10^3$	$9.9 \cdot 10$	$7.1 \cdot 10^3$
0.3	$5.7 \cdot 10$	$3.0 \cdot 10^3$	$6.8 \cdot 10$	$4.8 \cdot 10^3$

Table 33. The Numbers of Muons at Both Altitudes Whose Parental Pions Have Energies Larger than 110 Mc^{+2}

| Depth | E_0 | | | |
| | 720 gm cm^{-2} | | 1033 gm cm^{-2} | |
η	10^4	10^6	10^4	10^6
1	2.9 $(2.2 \cdot 10^{-2})$*	$6.1 \cdot 10$ $(7.2 \cdot 10^{-3})$	2.9 $(2.1 \cdot 10^{-2})$	$6.1 \cdot 10$ $(7.0 \cdot 10^{-3})$
0.5	$1.0 \cdot 10$ $(1.1 \cdot 10^{-2})$	$4.7 \cdot 10^2$ $(7.6 \cdot 10^{-2})$	$1.1 \cdot 10$ $(1.1 \cdot 10^{-1})$	$4.9 \cdot 10^2$ $(6.9 \cdot 10^{-2})$
0.3	$1.3 \cdot 10$ $(2.2 \cdot 10^{-1})$	$2.9 \cdot 10^2$ $(9.9 \cdot 10^{-2})$	$1.3 \cdot 10$ $(1.9 \cdot 10^{-1})$	$3.2 \cdot 10^2$ $(6.7 \cdot 10^{-2})$

* Values in the parentheses are ratios of the numbers of such muons to the numbers of total muons.

E_c has been chosen by Ueda and Ogita as 10 times the nucleon rest mass. The values of the other constants are given by $\alpha = 2$, $\gamma = 1.4$, and $\Delta = 0.7$. The collision mean free path has been chosen to be 90 grammes per square centimeter gm cm^{-2}.

On the basis of the above model, these authors have calculated the mean value of the different components of the extensive air shower. They have used the method of generation and traced the secondaries till they drop down below the critical energy. Tables 30–34 depict the mean values of the different components for various values of energy and depth.

Table 34. The Ratios of Various Combinations of the Numbers of
the Three Components

E_0		10^4			10^6		
η		1	0.5	0.3	1	0.5	0.3
$\dfrac{N+\mu}{\text{total no.}}$	M.L.*	$3.8\cdot10^{-2}$	$2.9\cdot10^{-2}$	$2.0\cdot10^{-2}$	$9.7\cdot10^{-3}$	$8.6\cdot10^{-3}$	$7.2\cdot10^{-3}$
	S.L.†	$3.3\cdot10^{-1}$	$1.5\cdot10^{-1}$	$6.1\cdot10^{-2}$	$5.9\cdot10^{-2}$	$2.9\cdot10^{-2}$	$1.6\cdot10^{-2}$
$\dfrac{\mu}{\text{total no.}}$	M.L.	$3.8\cdot10^{-2}$	$2.7\cdot10^{-2}$	$1.5\cdot10^{-2}$	$8.9\cdot10^{-3}$	$7.0\cdot10^{-3}$	$4.1\cdot10^{-3}$
	S.L.	$3.3\cdot10^{-1}$	$1.5\cdot10^{-1}$	$5.2\cdot10^{-2}$	$5.8\cdot10^{-2}$	$2.7\cdot10^{-2}$	$1.2\cdot10^{-2}$
$\dfrac{N}{N+\mu}$	M.L.	$9.2\cdot10^{-3}$	$8.4\cdot10^{-2}$	$2.6\cdot10^{-1}$	$7.1\cdot10^{-2}$	$1.8\cdot10^{-1}$	$4.3\cdot10^{-1}$
	S.L.	$6.4\cdot10^{-4}$	$1.7\cdot10^{-2}$	$1.4\cdot10^{-1}$	$9.0\cdot10^{-4}$	$5.9\cdot10^{-2}$	$2.7\cdot10^{-1}$

* M.L. means mountain level (720 gm cm^{-2}).
† S.L. means sea level (1033 gm cm^{-2}).

4. Size-Frequency Distribution of Showers

As we have observed in the introductory remarks, Heitler and Janossy [4] have estimated the size-frequency distribution of penetrating showers observed by Brown et al. [5]. We shall discuss the method of arriving at the probability that n mesons are emitted in a shower by a single nucleon. Let $\phi(E', E)\,dE'$ be the probability that a nucleon of energy E emits a meson and drops to an energy between E' and $E' + dE'$. The cross-section is assumed to be homogeneous. Thus ϕ is given by

$$(4.1)\quad \phi(E', E)\,dE' = \alpha\omega\left(\frac{E'}{E}\right)\frac{dE'}{E}, \qquad \int_0^E \omega\left(\frac{E'}{E}\right)\frac{dE'}{E} = 1$$

where α is the total cross-section. Equation (4.1) will hold good only for sufficiently high energies E, say above a critical energy E_c. We shall assume that

$$(4.2)\qquad\qquad \phi(E', E) = 0 \qquad (E < E_c).$$

Defining $\pi(n, E_0, E, x)\,dE$ as the joint probability that a nucleon of energy E_0 at $x = 0$ drops to an energy between E and $E + dE$ after traversing a thickness x and has undergone n collisions, we notice that π satisfies the equation

$$(4.3)\quad \frac{\partial\pi(n, E_0, E, x)}{\partial x} = -\alpha\,\pi(n, E_0, E, x)\,H(E - E_c)$$

$$+\alpha\int \pi(n - 1, E_0, E', x)\,\omega\left(\frac{E'}{E}\right)\frac{dE'}{E}.$$

It is convenient to break up $\pi(n, E_0, E, x)$ into two parts:

(4.4) $\qquad \pi^{(1)}(n, E_0, E, x) = \pi(n, E_0, E, x) \qquad (E < E_c)$
$\qquad\qquad \pi^{(2)}(n, E_0, E, x) = \pi(n, E_0, E, x) \qquad (E > E_c).$

$\pi^{(1)}$ and $\pi^{(2)}$ satisfy the equations

(4.5) $\qquad \dfrac{\partial \pi^{(2)}(n, E_0, E, x)}{\partial x} = -\alpha\, \pi^{(2)}(n, E_0, E, x)$

$$+ \alpha \int_E^{E_0} \pi^{(2)}(n-1, E_0, E', x)\omega\left(\frac{E'}{E}\right)\frac{dE'}{E}$$

(4.6) $\qquad \dfrac{\partial \pi^{(1)}(n, E_0, E, x)}{\partial x} = \alpha \int_{E_c}^{E_0} \pi^{(2)}(n-1, E_0, E', x)\,\omega\left(\frac{E'}{E}\right)\frac{dE'}{E}$

with the initial conditions

(4.7) $\qquad \pi^{(i)}(n, E_0, 0) = 0 \qquad (i = 1, 2) \qquad (n \neq 0)$
$\qquad\qquad \pi^{(1)}(0, E, 0) = 0$
$\qquad\qquad \pi^{(2)}(0, E, 0) = \left(\dfrac{E_c}{E}\right)^{\gamma+1}\dfrac{dE}{E_c}.$

Defining $p_n^{(1)}$ ($p_n^{(2)}$) as the probability that n mesons are emitted and that the energy of the primary is reduced to a value below (above) the critical energy, we find

(4.8) $\qquad \dfrac{\partial p_n^{(2)}(x)}{\partial x} = -p_n^{(2)}\alpha + \alpha\, p_{n-1}^{(2)}\,\alpha\omega_{\gamma+1}$

$\qquad\qquad \dfrac{\partial p_n^{(1)}(x)}{\partial x} = \alpha\, p_{n-1}^{(2)}(1 - \omega_{\gamma+1})$

where

$$\omega_\gamma = \int_0^E \omega\left(\frac{E'}{E}\right)\left(\frac{E'}{E}\right)^{\gamma+1}\frac{dE'}{E}.$$

The solutions of the equations are given by

(4.9) $\qquad p_n^{(2)}(x) = \dfrac{(\alpha x)^n}{n!}\, e^{-\alpha x}\,(\omega_{\gamma+1})^n$

$\qquad\qquad p_n^{(1)}(x) = \Gamma_n(\alpha x)\,\omega_{\gamma+1}^{n-1}(1 - \omega_{\gamma+1})$

where $\Gamma_n(a)$ is the incomplete gamma function defined by

$$(4.10) \qquad \Gamma_n(a) = \int_0^a \frac{z^{n-1}}{(n-1)!} \, e^{-z} \, dz.$$

Equation (4.9) is mainly the work of Heitler and Janossy [4]. However, the method of solution is due to Ramakrishnan [16].

We wish to observe that the above solution is valid only for the case *when there is a single primary whose energy is governed by right-hand side of (4.7).* However, experimentally it is the average spectrum of the type (4.7) that is measured. Thus the primary spectrum of nucleons defined by (4.7) is the product density of degree one and hence *is not a probability magnitude even though it has been normalized to unity.* The validity of equations (4.5) and (4.6) is questionable whenever we assume an energy spectrum for the primary nucleons. However, the first few moments of the number distribution can be determined by the method of product densities provided the initial spectrum is given in the form of correlations which, in turn, can be determined experimentally. Thus a detailed treatment based on product density technique analogous to the theory of development of electromagnetic cascades will provide a complete theory of extensive air showers.

REFERENCES

1. K. Schmeiser and W. Bothe, *Ann. Physik.*, **32** (1938), 161.
2. P. Auger, R. Maze, and T. Grivet-Meyer, *C.R. Acad. Sci.* (Paris), **206** (1938), 1721.
3. L. Janossy and A. C. B. Lowell, *Nature*, **142** (1938), 716.
4. W. Heitler and L. Janossy, *Proc. Phys. Soc.*, **62 A** (1949), 669.
5. R. H. Brown, U. Camerini, P. H. Fowler, H. Heitler, and C. P. Powell, *Phil. Mag.*, **40** (1949), 862.
6. H. Messel, *Prog. Cosm. Ray*, Vol. II (1952), Chapter IV.
7. K. Greisen, *Prog. Cosm. Ray*, Vol. III (1956), Chapter I.
8. H. Messel and R. B. Potts, *Nuovo Cimento*, **10** (1953), 754.
9. A. Ueda and N. Ogita, *Prog. Theor. Phys.*, **18** (1957), 269.
10. S. Olbert and R. Stora, *Ann. Phys.*, **1** (1957), 247.
11. H. Fukuda, N. Ogita, and A. Ueda, *Prog. Theor. Phys.*, **21** (1959), 29.
12. A. Ueda and C. B. A. McCusker, *Nucl. Phys.*, **26** (1961), 35.
13. A. Ueda, *Prog. Theor. Phys.*, **24** (1960), 1231.
14. H. Messel, *Proc. Phys. Soc.*, **64 A** (1951), 726.
15. A. Ramakrishnan and S. K. Srinivasan, *Prog. Theor. Phys.*, **11** (1954), 595.
16. A. Ramakrishnan, *Proc. Phys. Soc.*, **63 A** (1950), 861.

Chapter 8

POLARIZATION IN CASCADES

1. Introduction

The theory of electromagnetic cascades presented in Chapters 5 and 6 explains the relevant statistical properties of the shower particles with reference to their energies. However, there is one aspect of these showers that has not been taken into account. The electrons and photons have interesting polarization properties and their states of polarization undergo changes while traversing matter and experiencing collisions with the nucleii. Some calculations by Gluckstern, Hull, and Breit [1, 2] show that bremsstrahlung emitted at a particular angle from the direction of the incident electron is in general partially plane-polarized. In fact, this has been confirmed by experiments of Motz [3] and Dudley, Inman, and Kenney [4]. Hence it is reasonable to expect a certain degree of plane polarization among the photons in a cascade. Thus it is worthwhile to contemplate a theory of electromagnetic cascades incorporating all the aspects of the polarization properties of the particles involved. The problem in its very general form is indeed a formidable one since angular and lateral structure of the showers have to be studied in detail. However, there are one or two experimental results that enable us to deal with this problem at least from a limited viewpoint. The first is the exclusive longitudinal polarization of electrons resulting from muons (see, for example, Lee and Yang [5]). The other is the complete circular polarization of the high energy bremsstrahlung radiated by electrons originating in a beta-decay (see reference [6]). Thus it may be possible to formulate the theory of cascades initiated by a single electron or photon of controlled polarization. Motivated by these considerations, Dyson [7] has studied such cascades and obtained fairly good estimates for the mean behavior of longitudinally polarized particles. Ranganathan and Vasudevan [8] have extended these results on the basis of the new approach to cascade theory and have also outlined the method of estimating their size fluctuations. In this chapter, we shall give an account of these methods and also present some of the recent results of Srinivasan and Koteswara Rao [9] who have used the method of invariant imbedding technique and simplified the methods of calculation. Apart from this, there has been increasing interest in the polarization of cosmic ray muons [10, 11]. As the problem has not been formulated in correct theoretical terminology (see, for example, references [12, 13]), we have found it worthwhile to formulate the problem in terms of product densities and indicate the results that can be checked experimentally.

2. Product Densities of Polarized Particles

The inclusion of polarization brings in its wake a new feature hitherto not present in the cascade theory. An electron can exist in a "forward" (F) or "backward" (B) polarization pure state with respect to its original direction of motion or in any other state of polarization which can be expressed as a linear combination of the F and B states. This feature has to be included in the product density functions which are defined only in terms of energy. The description of "mixture states" is achieved by means of the density matrix method (see reference [14]) well known in quantum mechanics. To do this, we introduce the random variable $dN_{ij}(E, t)$ representing the number of particles each with energy in the range $(E, E + dE)$ and characterized by the polarization indices (i, j) (i and j take the discrete values F or B). Thus the random variable so introduced defines "product density matrices." The mean value of $dN_{FF}(E, t) - dN_{EE}(E, t)$ can be interpreted to be the product density magnitude of degree one of longitudinally polarized particles. In this sense, the product densities are defined over the product space of energy and polarization. An alternative mode of description consists of interpreting the product densities as density matrices thus justifying the terminology product density matrix.

2.1. MEAN BEHAVIOR

To describe the mean behavior of polarized particles we introduce the 2×2 density matrices $\pi_{ij}(E, t)$ and $\gamma_{ij}(E, t)$ where the labels i and j as before denote F and B signifying the forward and backward polarizations of the particles with respect to their directions of motion. We can interpret $\pi_{ij}(E, t)$ and $\gamma_{ij}(E, t)$ to be the mean number of electrons and photons with energies lying between E and $E + dE$ at the depth t with their states of polarization represented by the labels i and j.

The usual cascade diffusion equations involve the bremsstrahlung and pair production total cross-sections. When polarization is included in the cascade theory, we note that the progeny of a particle in the cascade depends very much on the state of polarization of the "parent." We now define the production matrix elements for the bremsstrahlung $M_{ikl}(E|E_0)$ where i, k, and l take

the values F and B representing respectively the spin states of the incident electron, the photon and the outgoing electron and E and E_0 denote the energies of
the photon and the incident electron. Similarly $N_{ikl}(E|E_0)$ is the pair production
matrix element where i, k, and l refer to the spin states of the incident photon,
the positron and the electron and E_0 and E are the energies of the photons
and the positrons.

Considering a bremsstrahlung process by an incident electron described
by a density matrix π_{ij}, the two outgoing particles can be described by the
4×4 joint density matrix

$$(2.1) \qquad J_{kl\,mn} = \sum_{ij} \pi_{ij}\, M_{ikl}\, M^\star_{jmn}.$$

Since we are interested only in the polarization of one of the outgoing
particles, we sum incoherently over the spin indices of the other particle and
thereby obtain the contracted (2×2) density matrix π'_{ln} given by

$$(2.2) \qquad \pi'_{ln} = \sum_k J_{kl\,mn} = \sum_{ijk} \pi_{ij}\, M_{ikl}\, M^\star_{jkn}.$$

For the outgoing photon, we have

$$(2.3) \qquad \gamma'_{ln} = \sum_l J_{kl\,ml} = \sum_{ijl} \pi_{ij}\, M_{ikl}\, M^\star_{jml}.$$

In a similar way, we can obtain the "contracted" density matrices for particles
resulting from pair production. The values of the different production
matrix elements in their relativistic limits can be calculated (see, for example,
reference [15]). Because of the peculiar angular dependence of these matrix
elements, on integration over angles the sum $\sum_k M_{lki}\, M^\star_{nkj}$ will always vanish
except when $l = n$, $i = j$, or $l = i$, $n = j$. Similarly the two sums $\sum_k M_{lik}\, M^\star_{njk}$
and $\sum_k N_{lki}\, N^\star_{nkj}$ are nonvanishing only in the case $l = n$, $i = j$; $\sigma_{ikl}(E|E_0)$
dE represents the cross-section integrated over angles for an electron of energy
E_0 to radiate photons in energy range $(E, E + dE)$ and $\tau_{ikl}(E|E_0)\, dE$ is the
cross-section for pair production. They are given by

$$(2.4) \qquad \sigma_{ikl}(E|E_0) = \int d\Omega\, |M_{ikl}(E|E_0)|^2$$
$$\tau_{ikl}(E|E_0) = \int d\Omega\, |N_{ikl}(E|E_0)|^2.$$

The diffusion equations for the product densities of degree one $\pi_{ij}(E|E_0, t)$
and $\gamma_{ij}(E|E_0, t)$ are given below:

$$(2.5) \qquad \frac{\partial}{\partial t} \pi_{ij}(E|E_0, t) = \int_0^{E_0} dE' \int d\Omega \sum_{lnk} \pi_{ln}\, (E'|E_0, t)\, M_{lki}\, M^\star_{nkj}$$

$$-\frac{1}{2}\int_{0}^{E} dE' \int d\Omega \sum_{kl} \pi_{ij}(E|E_0, t)\, [|M_{ikl}|^2 + |M_{jkl}|^2]$$

$$+2\int_{E}^{E_0} dE' \int d\Omega \sum_{lnk} \gamma_{ln}(E'|E_0, t)\, N_{lki}\, N_{nkj}^{\star}$$

$$(2.6)\quad \frac{\partial}{\partial t}\gamma_{ij}(E|E_0, t) = \int_{E}^{E_0} dE' \int d\Omega \sum_{lnk} \pi_{ln}(E'|E_0, t)\, M_{lik}\, M_{njk}^{\star}$$

$$-\frac{1}{2}\int_{0}^{E} dE' \int d\Omega \sum_{kl} \gamma_{ij}(E|E_0, t)[|N_{ikl}|^2 + |N_{jkl}|^2].$$

Using the properties of matrix elements under angular integration, we find that the above set of equations can be decoupled into four disjoint equations for the transverse polarization $(i \neq j)$ and two coupled equations for longitudinal polarization. The total population and the longitudinal components are given by

$$(2.7)\qquad \begin{aligned} \pi(E|E_0, t) &= \pi_{FF}(E|E_0, t) + \pi_{BB}(E|E_0, t) \\ \gamma(E|E_0, t) &= \gamma_{FF}(E|E_0, t) + \gamma_{BB}(E|E_0, t) \end{aligned}$$

$$(2.8)\qquad \begin{aligned} \pi_L(E|E_0, t) &= \pi_{FF}(E|E_0, t) - \pi_{BB}(E|E_0, t) \\ \gamma_L(E|E_0, t) &= \gamma_{FF}(E|E_0, t) - \gamma_{BB}(E|E_0, t). \end{aligned}$$

Mellin transform solutions for π_L and γ_L can be found using the methods outlined in Chapter 5 and these functions characterize the mean behavior of longitudinally polarized particles.

2.2. SECOND ORDER PRODUCT DENSITY MATRICES

To solve the fluctuation problem of the cascade theory with the inclusion of the state of polarization, it is obvious that we should define (4 × 4) density matrices $\pi_{ij,kl}(E_1, E_2, t)$ where (ij) and (kl) describe the states of polarization of particles 1 and 2 with energies lying between $(E_1, E_1 + dE_1)$ and $(E_2, E_2 + dE_2)$ respectively. Similar functions for photons and mixed functions for electrons and photons can be defined. The diffusion equations for these functions can easily be written down using essentially the same arguments for the equations of product densities of degree one.

To determine the fluctuation in longitudinal polarization we have to define four functions of the second degree for the electrons given by

$$(2.9) \quad \pi_{L_1 L_1}(E_1, E_2, t) = \pi_{FF,FF}(E_1, E_2, t) - \pi_{BB,BB}(E_1, E_2, t)$$
$$\pi_{L_1 L_2}(E_1, E_2, t) = \pi_{FF,BB}(E_1, E_2, t) - \pi_{BB,FF}(E_1, E_2, t)$$
$$\pi_{L_2 L_1}(E_1, E_2, t) = \pi_{BB,FF}(E_1, E_2, t) - \pi_{FF,BB}(E_1, E_2, t)$$
$$\pi_{L_2 L_2}(E_1, E_2, t) = \pi_{BB,BB}(E_1, E_2, t) - \pi_{FF,FF}(E_1, E_2, t)$$
$$= -\pi_{L_1 L_1}(E_1, E_2, t)$$

where L_1 and L_2 refer to the cases where an excess of electrons are polarized in the "forward" or "backward" directions respectively. Similarly, the longitudinal polarization functions for other product densities of degree two can be defined. The diffusion equations for these functions are related and can conveniently be cast in a vector-matrix form with some approximation (see reference [16]). A formal transform solution for the vector can be obtained and inverting this, we can recover the functions like $\pi_{L_1 L_1}(E_1, E_2; t)$.

3. Production Product Density Matrices of Longitudinally Polarized Particles

As we are interested in the longitudinally polarized particles created in the shower, we shall define suitable production product density matrices and obtain the required information using the invariant imbedding technique. In this case, however, we shall assume that the primary is in a general polarization state (ij). This is primarily motivated by the fact that even though we start with a longitudinally polarized particle, there is a non-vanishing probability for the particle to encounter a collision in $(0, \Delta)$ and consequently undergo a change in its polarization state. Let $f_1(E|E_0, ij, t)$ and $g_1(E|E_0, ij, t)$ be the production product density matrices of degree one and $f_2(E_1, E_2|E_0, ij, t_1, t_2)$ and $g_2(E_1, E_2|E_0, ij, t_1, t_2)$, the production product density matrices of degree two for longitudinally polarized electrons created respectively in the electron- and photon-initiated showers.

3.1. IMBEDDING EQUATIONS

Using the invariant imbedding technique we obtain the following differential equations satisfied by the first order product densities:

$$(3.1) \quad \frac{\partial f_1(E|E_0, ij, t)}{\partial t} = - \int_0^{E_0} dE' \int d\Omega \sum_{kl} f_1(E|E_0, ij, t) \tfrac{1}{2}(|M_{ikl}|^2 + |M_{jkl}|^2)$$

$$+ \int\limits_{0}^{E_0} dE' \int d\Omega \sum_{kln} [f_1(E|E', ln, t) M_{ikl} M^{\star}_{jkn} + g_1(E|E_0 - E', ln, t) M_{ilk} M^{\star}_{jnk}]$$

$$(3.2) \quad \frac{\partial g_1(E|E_0, ij, t)}{\partial t} = - \int dE' \int\limits_{0}^{E_0} d\Omega \sum_{kl} g_1(E|E_0, ij, t) \tfrac{1}{2}(|N_{ikl}|^2 + |N_{jkl}|^2)$$

$$+ 2 \int\limits_{E}^{E_0} dE' \int d\Omega \sum_{kln} f_1(E|E', lm, t) N_{ilk} N^{\star}_{jnk}.$$

The first term on the right-hand side of equation (3.1) arises when there occurs no event in $(0, \Delta)$ and the second and third terms are due to the occurrence of an event in $(0, \Delta)$. The second term governs the longitudinally polarized electrons produced in the shower initiated by the outgoing electron of energy E' and polarization state ln. As we are interested only in the shower initiated by the outgoing electron we incoherently sum over the spin states of the photon and thus obtain the contracted (2×2) density matrix. The third term is due to the shower initiated by the outgoing photon. In equation (3.2) we get a factor 2 in the second term on the right-hand side of the equation since we do not distinguish between the positron and electron in the theory of cascade showers. Here again we find the set of eight coupled equations given by (3.1) and (3.2) separate out, after the angular integrations are performed, into four uncoupled equations for transverse polarization, two coupled equations for longitudinal polarization, and two coupled equations for total population of the primaries.

Defining the functions $f_1(E|E_0, L, t)$ and $g_1(E|E_0, L, t)$ (analogous to π_L and γ_L defined earlier) for the longitudinally polarized electron- and photon-initiated showers, by

$$(3.3) \quad \begin{aligned} f_1(E|E_0, L, t) &= f_1(E|E_0, FF, t) - f_1(E|E_0, BB, t) \\ g_1(E|E_0, L, t) &= g_1(E|E_0, FF, t) - g_1(E|E_0, BB, t) \end{aligned}$$

we deduce from (3.1) and (3.2) the equations satisfied by them. Since the polarization cross-sections are dependent only on the ratio of the energies of the initial electron and the outgoing photon (initial photon and the positron produced) we can write $f_1(E|E_0, L, t) dE$ and $g_1(E|E_0, L, t) dE$ respectively as $f_1(\mathscr{E}, L, t) d\mathscr{E}$ and $g_1(\mathscr{E}, L, t) d\mathscr{E}(\mathscr{E} = E/E_0)$. Defining the Mellin transform of $f_1(\mathscr{E}, L, t)$ and $g_1(\mathscr{E}, L, t)$ with respect to $\mathscr{E}$ as $f_1(s, L, t)$ and $g_1(s, L, t)$, we obtain

(3.4)
$$\frac{\partial}{\partial t}\begin{bmatrix} f_1(s, L, t) \\ g_1(s, L, t) \end{bmatrix} = \begin{bmatrix} -A_L(s) & C_L(s) \\ B_L(s) & -D \end{bmatrix}\begin{bmatrix} f_1(s, L, t) \\ g_1(s, L, t) \end{bmatrix}$$

together with the initial conditions

(3.5)
$$\begin{bmatrix} f_1(s, L, 0) \\ g_1(s, L, 0) \end{bmatrix} = \begin{bmatrix} 0 \\ B_L(s) \end{bmatrix}$$

where

(3.6)
$$A_L(s) = \left(\frac{4}{3} + \alpha\right)[\psi(s) - \psi(1)] + \frac{1}{2} - \frac{\left(\frac{1}{3} + \alpha\right)}{s(s+1)}$$

$$B_L(s) = \frac{2}{s} - \frac{2}{s(s+1)}\left(\frac{4}{3} + \alpha\right).$$

$$C_L(s) = \frac{1}{s+1} + \left(\frac{4}{3} + \alpha\right)\frac{1}{s(s+1)}$$

$$D = \frac{7}{9} - \frac{\alpha}{6}.$$

The mean number of longitudinally polarized electrons created in a shower initiated by a longitudinally polarized electron (photon) in thickness $(0, t)$ above an energy E is given by

(3.7)
$$N(\mathscr{E}, t) = \frac{1}{2\pi i} \int_{\sigma-i\infty}^{\sigma+i\infty} \frac{\mathscr{E}^{1-s} B_L(s) C_L(s)}{(s-1)[\mu_L(s) - \lambda_L(s)]}$$

$$\left[\frac{1 - e^{-\lambda_L(s)t}}{\lambda_L(s)} - \frac{1 - e^{-\mu_L(s)t}}{\mu_L(s)}\right]ds$$

(3.8)
$$M(\mathscr{E}, t) = \frac{1}{2\pi i} \int_{\sigma-i\infty}^{\sigma+i\infty} \frac{B_L(s)\,\mathscr{E}^{1-s}}{(s-1)[\mu_L(s) - \lambda_L(s)]}$$

$$\left[\frac{D - \lambda_L(s)}{\mu_L(s)}(1 - e^{-\mu_L(s)t}) - \frac{D - \mu_L(s)}{\lambda_L(s)}(1 - e^{-\lambda_L(s)t})\right]ds$$

where $-\lambda_L(s)$ and $-\mu_L(s)$ are the eigen values of the matrix

$$\begin{bmatrix} -A_L(s) & C_L(s) \\ B_L(s) & -D \end{bmatrix}.$$

3.2. EXPLICIT SOLUTION OF THE MEAN NUMBERS

The mean numbers of logitudinally polarized electrons created in the entire shower are given by

(3.9)
$$N(\mathscr{E}) = \frac{1}{2\pi i} \int_{\sigma-i\infty}^{\sigma+i\infty} \frac{e^{y(s-1)}}{(s-1)} \frac{B_L(s)\, C_L(s)}{A_L(s)\, D - B_L(s)\, C_L(s)}\, ds$$

$$M(\mathscr{E}) = \frac{1}{2\pi i} \int_{\sigma-i\infty}^{\sigma+i\infty} \frac{e^{y(s-1)}}{s-1} \frac{B_L(s)\, A_L(s)}{A_L(s)\, D - B_L(s)\, C_L(s)}\, ds$$

where $y = - \log \mathscr{E}$ and $N(\mathscr{E})$ and $M(\mathscr{E})$ denote the limit of $N(\mathscr{E}, t)$ and $M(\mathscr{E}, t)$ respectively as t tends to infinity. To evaluate the contour integrals occurring in equations (3.9), we note that the integrand in either of the two equations in (3.9) does not have any singularity to the right of $s = 2$ in the complex s-plane. At $s = 0.313$ and 1.368 there are simple poles due to the zeros of $\phi_L(s) = A_L(s)\, D - B_L(s)\, C_L(s)$, on the portion of the real axis extending from -1 to 0.

An analysis similar to the one indicated in Section 4 of Chapter 6 shows that $\phi_L(s)$ has two zeros off the real axis. Choosing $\sigma = \sigma_0 > 1.5$ we can evaluate the Mellin integral. Following the procedure adopted in Section 4 of Chapter 6 we can show that $N(\mathscr{E})$ can be approximated by the sum of the residues at the poles at $s = 1, 0.313,$ and 1.368 and possibly at the two zeros off the real axis. In the light of our experience in Chapter 6, we omit the contributions from these zeros and obtain the following analytical expressions for $N(\mathscr{E})$ and $M(\mathscr{E})$:

(3.10)
$$N(\mathscr{E}) = 1.2446\, \mathscr{E}^{-0.368} - 0.0576\, \mathscr{E}^{0.687} - 1.4884$$
$$M(\mathscr{E}) = 1.0049\, \mathscr{E}^{-0.368} + 0.0164\, \mathscr{E}^{0.687} - 0.4052.$$

This approximation will be very good for large y. For small values of y we may have to incorporate the contributions from the first few poles that are encountered as we move to the left from zero.

3.3. HIGHER MOMENTS

In order to evaluate the second moments of the number distribution of longitudinally polarized electrons we shall write the invariant imbedding equations satisfied by the second order product densities

$$f_2(E_1, E_2 | E_0, ij, t_1, t_2) \quad \text{and} \quad g_2(E_1, E_2 | E_0, ij, t_1, t_2).$$

They are given by

(3.11a) $\left(\dfrac{\partial}{\partial t_1} + \dfrac{\partial}{\partial t_2}\right) f_2(E_1, E_2 | E_0, ij, t_1, t_2)$

$$= - \int_0^{E_0} dE' \int d\Omega \sum_{kl} \frac{1}{2}(|M_{ikl}|^2 + |M_{jkl}|^2) f_2(E_1, E_2 | E_0, ij, t_1, t_2)$$

$$+ \int_{E_1}^{E_0} dE' \int d\Omega \sum_{lnk} [f_2(E_1, E_2 | E', ln, t_1, t_2) M_{ikl} M_{jkn}^{\star}$$

$$+ g_2(E_1, E_2 | E_0 - E', ln, t_1, t_2) M_{ilk} M_{jnk}^{\star}]$$

$$+ \int_{E_1}^{E_0} dE' \int d\Omega \sum_{klmn} [f_1(E_1 | E', ln, t_1) g_1(E_2 | E_0 - E', km, t_2)$$

$$+ f_1(E_2 | E', ln, t_2) g_1(E_1 | E_0 - E', km, t_1)] M_{ikl} M_{jmn}^{\star},$$

(3.11b) $\left(\dfrac{\partial}{\partial t_1} + \dfrac{\partial}{\partial t_2}\right) g_2(E_1, E_2 | E_0, ij, t_1, t_2)$

$$= - \int_0^{E_0} dE' \int d\Omega \sum_{kl} \frac{1}{2}(|N_{ikl}|^2 + |N_{jkl}|^2) g_2(E_1, E_2 | E_0, ij, t_1, t_2)$$

$$+ \int dE' \int d\Omega \sum_{lnk} [f_2(E_1, E_2 | E', ln, t_1, t_2) N_{ikl} N_{jkn}^{\star}$$

$$+ f_2(E_1, E_2 | E_0 - E', ln, t_1, t_2) N_{ilk} N_{jnk}^{\star}]$$

$$+ 2 \int dE' \int d\Omega \sum_{klmn} f_1(E_1 | E', ln, t_1) f_1(E_1 | E_0 - E', km, t_2) N_{ikj} N_{jmn}^{\star}.$$

Defining

(3.12) $\qquad f_2(E_1, E_2 | E_0, L, t_1, t_2) = f_2(E_1, E_2 | E_0, FF, t_1, t_2)$

$$- f_2(E_1, E_2 | E_0, BB, t_1, t_2)$$

$\qquad g_2(E_1, E_2 | E_0, L, t_1, t_2) = g_2(E_1, E_2 | E_0, FF, t_1, t_2)$

$$- g_2(E_1, E_2 | E_0, BB, t_1, t_2)$$

we can deduce the equations satisfied by these new functions. In obtaining those equations we shall make use of the fact that longitudinally polarized electrons cannot be produced in a shower initiated by a transverse polarized electron or photon. (This can be easily established by solving equations (3.1) and (3.2) with $i \neq j$ under the initial conditions

$$f_1(E | E_0, ij, 0) = 0, \quad g_1(E | E_0, ij, 0) = 0) \qquad (i \neq j)$$

Again writing $f_2(E_1, E_2 | E_0, L, t_1, t_2)\, dE_1\, dE_2$ and $g_2(E_1, E_2 | E_0, L, t_1, t_2)\, dE_1\, dE_2$ as $f_2(\mathscr{E}_1, \mathscr{E}_2, L, t_1, t_2)\, d\mathscr{E}_1\, d\mathscr{E}_2$ and $g_2(\mathscr{E}_1, \mathscr{E}_2, L, t_1, t_2)\, d\mathscr{E}_1\, d\mathscr{E}_2$ where $\mathscr{E}_1 = E_1/E_0,\ \mathscr{E}_2 = E_2/E_0$ and defining

$$(3.13)\quad f_2(s, L) = \lim_{t \to \infty} \int_0^1 \mathscr{E}^{s-1}\, d\mathscr{E} \int_{\mathscr{E}}^1 d\mathscr{E}_1 \int_{\mathscr{E}}^1 d\mathscr{E}_2 \int_0^t \int_0^t f_2(\mathscr{E}_1, \mathscr{E}_2, L, t_1, t_2)\, dt_1\, dt_2$$

$$g_2(s, L) = \lim_{t \to \infty} \int_0^1 \mathscr{E}^{s-1}\, d\mathscr{E} \int_{\mathscr{E}}^1 d\mathscr{E}_1 \int_{\mathscr{E}}^1 d\mathscr{E}_2 \int_0^t \int_0^t g_2(\mathscr{E}_1, \mathscr{E}_2, L, t_1, t_2)\, dt_1\, dt_2$$

we obtain after some calculation

$$(3.14)\qquad f_2(s, L) = \frac{D\, Q_1(s+1) + C_L(s+1)\, Q_2(s+1)}{A_L(s+1)\, D - B_L(s+1)\, C_L(s+1)}$$

$$g_2(s, L) = \frac{B_L(s+1)\, Q_1(s+1) + A_L(s+1)\, Q_2(s+1)}{A_L(s+1)\, D - B_L(s+1)\, C_L(s+1)}$$

where $Q_1(s)$ and $Q_2(s)$ are functions of s similar to $L_1(s)$ and $L_2(s)$ defined by (4.17) and (4.18) of Chapter 6. They are, in fact, functionals of $R_1(\mathscr{E})$ and $R_2(\mathscr{E})$ and have been evaluated in reference [9]. We refrain from giving their explicit forms as they are unduly lengthy. We observe that $f_2(s, L)$ and $g_2(s, L)$ are the Mellin transforms of the second factorial moment of the number of longitudinally polarized electrons generated by a longitudinally polarized primary. To find the inverses of $f_2(s, L)$ we can use the method indicated in Section 4 of Chapter 6. However, the expressions for the mean square do not take a simple form. It may be worthwhile to examine whether a direct numerical computation of $f_2(\mathscr{E}, L)$ and $g_2(\mathscr{E}, L)$ following the method of Bellman *et al.* [17] is feasible.

In a similar manner, we can obtain the higher moments of the polarized particles following the methods outlined in Section 4.3 of Chapter 6.

4. Polarization of Cosmic Ray Muons

Ever since the report on measurements of polarization of cosmic ray muons by Clark and Hersil [10], a good deal of interest has been evinced in this aspect of cosmic rays [17–20]. As the polarization of muons depends in part on the relative numbers of kaons and pions (which are the main sources of cosmic ray muons), polarization measurements provide us with valuable information on the production of kaons and pions at cosmic ray energies. The object of the present section is to propose a theoretical method by which we can calculate results that can be compared with the experiments. The work of Berezinskii and Dolgoshein [12, 13] served as the main stimulus for the present method of approach.

4.1. SEQUENT CORRELATION FUNCTIONS

The experimental quantity that is measured is the polarization of muons that are produced at different heights and have an energy E at sea level. Further, owing to ionization losses, the energy at the point of production of muons increases with the heights and it is this energy that more or less determines the degree of polarization.* Thus we have to deal with the muons that are observed at sea level with reference to the energy and the polarization. The quantity that can be defined in such a process is the number of muons having an energy E in a given interval and the polarization characterized by a certain small cone whose angle is the angle between the muon spin and momentum direction. More precisely, it is the probability distribution of the number of muons that is meaningful from a theoretical point of view. Thus we can define $\pi(n, E, \theta)$ as the probability that n muons (at sea level) have an energy above E, the polarization of each of the muons being characterized by θ. The difficulty of dealing with such a π-function is well known at least in other contexts. However, if we confine ourselves to the moments of the π function, it is possible to obtain some results by the use of product densities. In fact, a satisfactory description of the muons in the present context can be given by the use of sequent correlation functions developed in Chapter 4. Thus we can define $f_1(E_1, a, t_1; E_2, t)$ as the corresponding sequent correlation functions where $f_1(E_1, a, t_1; E_2, t) \, dE_1 \, dE_2 \, dt_1 \, da$ denotes the probability that a muon is produced between t_1 and $t_1 + dt_1$ with an energy in the range $(E_1, E_1 + dE_1)$ and the polarization in the range $(a, a + da)$ and is observed at t with energy in the range $(E_2, E_2 + dE_2)$. According to the theory of product densities, the expression

$$\int_0^t dt_1 \int_E^\infty dE_2 \int_{E_2}^\infty f_1(E_1, a, t_1; E_2, t) \, dE_1$$

represents the mean number of muons that have an energy above E at sea level and polarization characterized by a. Thus the information on the number spectrum of muons can be completely obtained with the help of the sequent correlation function of order one. If, however, fluctuation about the mean

---

* We neglect depolarization effects of the muon in its sojourn through the atmosphere since such effects do not exceed a few percent. See, for example, the report by T. L. Asatiani, *Proc. Int. Conf. Cosmic Rays*, Jaipur (1963).

number is required, the sequent correlation function of order two defined by
$f_2(E_1, a_1, t_1, E_2, a_2, t_2; E_3, E_4, t)$ where $f_2\, dE_1\, dE_2\, da_1\, da_2\, dE_3\, dE_4\, dt_1\, dt_2$
represents the joint probability of finding two muons at t (corresponding to
sea level) with energies respectively in the range $(E_3, E_3 + dE_3)$ and
$(E_4, E_4 + dE_4)$, the muons having been produced with energies in the interval
$(E_1, E_1 + dE_1)$, $(E_2, E_2 + dE_2)$ and polarizations in the intervals $(a_1, a_1 + da_1)$ and $(a_2, a_2 + da_2)$.

The integral of f_2 over $E_1, E_2, t_1, t_2, a_1, a_2, E_3, E_4$ yields the second factorial
moment of the number distribution. The sequent correlation function of
order one can be computed if we know the energy spectrum of pions at any
depth t.

4.2. PION SPECTRUM

To determine the pion π-meson spectrum following Berezinskii [12] we
can deal with the diffusion equation

$$(4.1) \qquad \frac{\partial \pi(\varepsilon, x)}{\partial x} = -\chi(E)\pi(\varepsilon, x) + B_0\, e^{-\mu x}\, \varepsilon^{-\gamma} + \frac{E\pi}{x\varepsilon}\, \pi\, (\varepsilon, x)$$

governing the vertical component of the π-meson beam, where $\pi(\varepsilon, x)$ is the
number of π-mesons with energy ε at the atmospheric depth x in nuclear
units; we have neglected the ionization losses suffered by the pions.

The first term on the right-hand side of equation (4.1) describes the absorption of π-mesons in the atmosphere as a result of nuclear interactions; $\chi(\varepsilon)$
is the inverse of the mean effective range for inelastic collisions of π-mesons
with air nuclei The second term represents the production of π-mesons by
the nucleonic component which is absorbed in the atmosphere exponentially;
$\mu = 0.65$ is the inverse of the mean free path of primary protons, measured
in nuclear lengths. The energy spectrum of the π-meson production in the
energy range under consideration can be assumed to follow a power law
with exponent $\gamma = 2.65$. The form of the term describing the π-meson
production renders the equation invalid for energies $\varepsilon < 2$ BeV.

The third term in equation (4.1) describes the π-meson decay; (E_π/xE) is
the decay probability of a π-meson with energy E traversing one nuclear
length; $E_\pi = 10^{11}$ eV.

The solution of equation (4.1) has been obtained by Berezinskii under the
boundary condition $\pi(E, x) = 0$ when $x = 0$ as

$$(4.2) \qquad \pi(\varepsilon, x) = B_0\, e^{-\mu x}\, \frac{\varepsilon^{-\gamma}\, x}{1 + E\pi/\varepsilon}$$

$$\left[1 - \frac{ax}{2 + E\pi/\varepsilon} + \frac{a^2 x^2}{(2 + E\pi/\varepsilon)(3 + E\pi/\varepsilon)} - \cdots \right].$$

Taking into account that at an altitude x, $(m\pi/\varepsilon\tau_0)\,\pi(\varepsilon, x)\,dx$ pions with energy ε decay per unit time (τ_0 being the mean lifetime in its rest frame) Berezinskii obtained an explicit expression for $N(E, \varepsilon, x)$, the mean number of μ-mesons (or mu-mesons) with energy E produced in the decay of π-mesons with energy ε at the altitude x, as

(4.3) $N(E, \varepsilon, x) = Bx\, e^{-\mu x} f(E, \varepsilon, x)$

(4.4) $f(E, \varepsilon, x) = \dfrac{\varepsilon^{-(2+\gamma)}}{(1 + E\pi/\varepsilon)\sqrt{1 - (m\pi/\varepsilon)^2}}$

$$\left[1 - \frac{ax}{2 + E\pi/\varepsilon} + \frac{a^2x^2}{(2 + E\pi/\varepsilon)(3 + E\pi/\varepsilon)} - \cdots\right]$$

where B is a new constant. Mu-mesons with energy E are produced from π-mesons with energy ε in the range $\varepsilon_- < \varepsilon < \varepsilon_+$ where

(4.5) $\varepsilon_+ = (EE^* + pp^*)\,m_\pi/m_\mu{}^2, \quad \varepsilon_- = (EE^* - pp^*)m_\pi/m_\mu{}^2.$

The polarization of cosmic ray μ-mesons has been calculated by Hayakawa [21], who, by means of relativistic transformations of the spin 4-pseudo-vector, found the polarization of μ-mesons with energy E produced in the decay of π-mesons with energy ε in the laboratory system of coordinates:

(4.6) $\cos\theta = \dfrac{EE^*}{pp^*} = \dfrac{\varepsilon m_\mu^2}{pp^* m_\pi}$

$E^* = (m_\pi^2 + m_\mu^2)/2m_\pi, \quad p^* = (m_\pi^2 - m_\mu^2)/2m_\pi$

where θ is the angle between the μ-meson spin and its momentum.

4.3.   NUMBER SPECTRUM OF MUONS

We note that equation (4.3) can be identified as the product density of degree one (which is nothing but the mean number spectrum) of pions. With the help of this expression, we can arrive at $f_1(E_1, a_1, t_1; E_2, t)$ by using the Hayakawa formula (4.6) for the in-flight polarization of a muon in pi-mu decay. *However, the expression obtained by weighting the right-hand side of (4.3) by cos θ cannot be identified to be the mean polarization.* This is due to the fact that the product density (which can be identified to be the differential mean spectrum) cannot be identified with the probability density since the integral of the former will always yield the mean number available in the range of integration of the parameters (here the parameters stand for energy and polarization). The authors of references [12] and [13] have apparently taken the product density of the number of muons to be unnormalized probability density and hence obtained the mean polarization by weighting the Hayakawa's expression for $\cos \theta$ (θ is the angle between the muon spin and momentum direction) by the corresponding normalized product density. In our opinion, this not a correct procedure and as a matter of fact, no such quantity as average polarization can be obtained in so simple a manner. In fact, in terms of the f_1 function defined, average value of the sum of $\cos \theta$ values of all the muons observed at sea level is given by

$$\int dE_1 \int dt_1 \int da_1 \int dE_2 \quad f_1(E_1, a_1, t_1; E_2, t) \cos \theta$$

and equation (9) of reference [13] for the mean polarization is nothing but the ratio of the above expression to the mean number of muons. To be more explicit, if n is the number of muons and θ_i is the angle corresponding to the i-th muon, mean polarization should be the mean value of

$$\frac{1}{n} \Sigma \cos \theta_i$$

and this is obviously not equal to

$$\overline{\left[\sum_{i=1}^{n} \cos \theta_i \right]} \Big/ \bar{n}.$$

It is our contention that the results on muon polarization as presented in references [12], [13], and [22] should be completely re-expressed in terms of the sequent correlation functions as explained above.

REFERENCES

1. R. L. Gluckstern, M. H. Hull, and G. Breit, *Phys. Rev.*, **90** (1953), 1026.
2. R. L. Gluckstern and M. H. Hull, *Phys. Rev.*, **90** (1953), 1030.
3. J. W. Motz, *Phys. Rev.* **104** (1956), 557.
4. J. M. Dudley, F. W. Inman, and R. W. Kenney, *Phys. Rev.*, **102** (1956), 925.
5. T. D. Lee and C. N. Yang, *Phys. Rev.* **105** (1957), 1671.
6. M. Goldhaber, L. Grodzins, and A. W. Sunyar, *Phys. Rev.* **106** (1957), 826.
7. F. J. Dyson, *Polarisation in Cascades*, Institute for Advanced Study Preprint (1957).
8. N. R. Ranganathan and R. Vasudevan, *Proc. Phys. Soc.*, **76** (1960), 650.
9. S. K. Srinivasan and N. V. Koteswara Rao (1969) (to be published).
10. G. W. Clark and J. Hersil, *Phys. Rev.*, **108** (1957), 1538.
11. H. V. Bradt and G. W. Clark, *Phys. Rev.*, **132** (1963), 1306.
12. V. S. Berezinskii, *Soviet Phys.—JETP* **15** (1962), 340.
13. V. Berezinskii and B. Dolgoshein, *Soviet Phys.—JETP*, **15** (1962), 749.

14. U. Fano, *Rev. Mod. Phys.*, **26** (1957), 70.
15. K. W. Mc Voy and F. J. Dyson, *Phys. Rev.*, **106** (1957), 1360.
16. R. Vasudevan and N. R. Ranganathan, *Proc. of the Summer School of Theoretical Physics* (Mussorie), Min. of Sci. Res. and Cultural Affairs (Govt. of India), New Delhi, Part II (1959), 261.
17. R. E. Bellman, R. E. Kalaba, and J. Lockett, *Numerical Inversion of the Laplace Transform*, American Elsevier Publishing Company, New York, 1966.
18. J. M. Fowler, H. Primakoff, and R. D. Sard, *Phys. Rev.*, **108** (1957), 1538.
19. B. Dolgoshein and B. I. Luchkov, *Soviet Phys.—JETP*, **9** (1959), 445.
20. B. Dolgoshein, B. I. Luchkov and V. Ushakov, *Soviet Phys.—JETP*, **15** (1962), 654.
21. S. I. Hayakawa, *Phys. Rev.*, **108** (1957), 1533.
22. B. Berezinskii, *Soviet Phys.—JETP*, **16** (1963), 657.

Chapter 9

POPULATION GROWTH

1. Introduction

The stochastic problem of population growth has attracted the attention of the probability theorists ever since 1874 when Watson and Galton [1] formulated the problem of extinction of families. In the forties of the present century, the problem received further impetus partly because of the increased general interest in the theory of stochastic processes and its applications and partly because of the analogy between the population growth and other multiplicative processes occurring in the realms of nuclear reactions and of cosmic rays. In fact, in the symposium on stochastic processes arranged by the Royal Statistical Society in June 1949 [2] a considerable part of the proceedings dealt with some aspect or other of the general problem of multiplicative processes. In a fundamental paper presented at the symposium, Kendall [3] has dealt with the problem of age specific population growth by means of the cumulant generating functional and its functional derivatives. These functional derivatives are nothing but the cumulant functions discussed in Chapter 2 and are very closely related to the product densities of Ramakrishnan dealt with in Chapter 3 of this monograph. Population growth has been studied intensively since then. The object of the present chapter is to present one particular aspect that has not been covered so far—the product density approach.

The layout of the present chapter is as follows. Section 2 will deal with the fluctuation in the population size on the basis of an age dependent model due to Kendall [3]. We have taken the opportunity to discuss some of the unpublished work of Ramakrishnan (see, Kendall [3]) who has derived all the results of Kendall by the product density method. We next discuss the method of invariant imbedding technique and a direct method of evaluating the product densities. In the final sections, some models of carcinogenesis are discussed.

2. Age Dependent Birth and Death Processes

Kendall [3] formulated the problem of population growth in the following

190

manner. The given primary of age x_0 at time $t = 0$ and its subsequent secondaries generate a population in accordance with the assumptions:

(i) the sub-populations generated by two coexisting individuals develop in complete independence of one another;

(ii) an individual of age x at time t has a probability

$$\lambda(x)\, dt + o(dt)$$

of producing a single individual of age zero in the interval $(t, t + dt)$;

(iii) an individual of age x existing at time t has a probability

$$\mu(x)\, dt + o(dt)$$

of death in the interval $(t, t + dt)$;

(iv) the birth and death probabilities $\lambda(x)$ and $\mu(x)$ depend only on the age x of the individual and not on t, the time of its existence.

We wish to obtain the probability frequency function of the number of individuals above a certain age at time t. We will see presently that this is a very difficult problem under the most general conditions and explicit solution for the probability frequency function is possible only under very special circumstances.

2.1. PRODUCT DENSITIES—INTEGRAL EQUATIONS

Let $f_1(x|x_0; t)$ and $f_2(x, y|x_0; t)$ be the product densities of degree one and two of individuals at time t generated by a primary of age x_0 at $t = 0$ so that $f_1(x|x_0; t)\, dx$ denotes the probability that there exists an "individual" of age between x and $x + dx$ at time t and that $f_2(x, y|x_0; t)\, / dx\, dy$ denotes the probability that there exist an individual of age between x and $x + dx$ and an individual of age between y and $y + dy$ at time t. Then $f_1(x|x_0; t)$ satisfies the equation*

$$(2.1)\quad f_1(x|x_0; t) = e^{-\mu t}\, \lambda(x_0 + t - x) + e^{-\mu x} \int_0^{t-x} f_1(u|x_0; t - x)\lambda(u)\, du.$$

The above equation is obtained by observing that in order that an individual of age between x and $x + dx$ exists at time t, it should have been born between $t - x - dx$ and $t - x$ and that the birth may be caused either by the primary or by a secondary.

We next observe that for fixed x_0, $f_1(x|x_0; t)\, e^{\mu t}$ is a function of $t - x$ only.

* Throughout we shall take $\mu(x) = $ a constant $= \mu$.

Setting

$$(2.2) \qquad\qquad f_1(x|x_0; t)\, e^{\mu x} = \phi(t - x)$$

we obtain

$$(2.3) \qquad \phi(u) = \lambda(x_0 + u)\, e^{-\mu u} + \int_0^u \lambda(y)\, e^{-\mu y}\, \phi(u - y)\, dy.$$

Kendall has proved the existence and uniqueness of the solution of equation (2.3).

To obtain an integral equation satisfied by $f_2(x, y|x_0; t)$, we observe that f_2 satisfies the conditions

$$(2.4) \qquad\qquad f_2(x, y|x_0; t) = f_2(y, x|x_0; t)$$

$$(2.5) \qquad\qquad f_2(x, y|x_0; t) = 0 \qquad\qquad x > t \text{ or } y > t$$

since we deal only with secondaries. Therefore, it is enough if we determine $f_2(x, y; t)$ for $x < y < t$. To obtain an equation for $f_2(x, y; t)$ we note that at time $t - x - dx$, there should be an individual of age between $y - x - dx$ and $y - x$ and that an individual must be born in the infinitesimal interval $(t - x - dx, t - x)$. In this case, we note that the birth can be caused by the primary or by the secondary of age between $y - x - dx$ and $y - x$ or by any other secondary. By weighting with correct probability functions, we obtain

$$(2.6) \qquad f_2(x, y|x_0; t) = \lambda(x_0 + t - x)\ \ k(y - x, t - x)\, e^{-2\mu x}$$
$$+ \lambda(y - x) f_1(y - x|x_0, t - x)\, e^{-2\mu x}$$
$$+ \int_0^{t-x} f_2(y - x, u|x_0; t - x)\, \lambda(u)\, e^{-2\mu x}$$

where the subsidiary function $K(u, v)$ has the following interpretation: $K(u, v)\, du$ denotes the joint probability that the primary is alive and that an individual of age between u and $u + du$ is found at time v. K satisfies the equation

$$(2.7) \quad K(x, t) = \lambda(x_0 + t - x)\, e^{-\mu(t-x)\ -2\mu x} + e^{-2\mu x} \int_0^{t-x} \lambda(y)\, K(y, t - x)\, dy.$$

We again note that $K(x, t) e^{\mu(t+x)}$ is a function of $t - x$ so that if we set

(2.8) $$K(x, t) = e^{-\mu(t+x)} \chi(t - x)$$

we obtain

(2.9) $$\chi(u) = \lambda(x_0 + u) + \int_0^u \lambda(y)\chi(u - y) e^{-\mu y} dy.$$

The existence and uniqueness of $\chi(u)$ has been established by Kendall. Equation (2.6) is reduced to a simpler form by observing that for $x < y < t$ and constant x_0,

(2.10) $$f_2(x, y|x_0; t) = e^{-2\mu t} \psi(y - x, t - y)$$

so that ψ satisfies the equation

(2.11) $\psi(u, v) = \lambda(x_0 + u + v) e^{2\mu(u+v)} K(u, u + v) + \lambda(u) e^{2\mu v + \mu u} \phi(v)$

$$+ \int_0^u \psi(u - u', v)\lambda(u') \, du' + \int_0^v e^{2\mu(u+u')} \psi(u', v)\lambda(u + u') \, du'.$$

Equation (2.11) is a new type of integral equation and has not been investigated analytically.

When the birth rates are not age-specific, equations (2.3), (2.9), and (2.11) can be solved explicitly.

(2.12) $$f_1(x|x_0; t) = \lambda e^{(\lambda - \mu)t - \lambda x}$$

(2.13) $$K(x, t) = \frac{\lambda}{\lambda - \mu} e^{-\mu(t+x)} (\lambda e^{(\lambda - \mu)(t-x)} - \mu)$$

(2.14) $$f_2(x, y|x_0; t) = \frac{\lambda^2}{\lambda - \mu} e^{-2\mu t + \lambda(t-y)} [2\lambda e^{\lambda(t-x)}$$

$$- \mu e^{\lambda(y-x)+\mu(t-y)} - \mu e^{\mu(t-x)}].$$

However, when the birth rates are age-specific, the integral equations appear to be intractable. We shall show in Section 3 that it is possible to write partial differential equations by the principle of invariant imbedding and obtain explicit solutions at least for some special forms of $\lambda(x)$.

2.2. MULTIPLE PRODUCT DENSITIES

Kendall's model of birth and death process has been generalized by Ramakrishnan and Srinivasan [4] who have introduced the following assumptions:

(i) an individual of age x existing at time t has a probability

$$\lambda_2(x)\, dt + o(dt)$$

of producing a twin of age zero between t and $t + dt$, $\lambda_2(x)$ being independent of t, and

(ii) the two individuals constituting a twin behave individually after production.

In view of these assumptions, we have a probability $O(dx)$ of finding two individuals in the same age group $(x, x + dx)$ at any particular time. In the notation of Chapter 4, if we use $dM(x, t)$ to denote the total number of individuals in the age group $(x, x + dx)$ at time t and $dN_1(x, t)$ and $dN_2(x, t)$ to denote the number of singles and twins respectively, we have the relation

(2.15) $$dM(x, t) = dN_1(x, t) + 2\, dN_2(x, t).$$

It is to be noted that $dM(x, t)$ is no longer asymptotically Poisson while $dN_1(x, t)$ and $dN_2(x, t)$ are. The variance of $dM(x, t)$ is given by

(2.16) $$\text{var } dM(x, t) = \text{var } dN_1(x, t) + 4 \text{ var } dN_2(x, t)$$

which is greater than the mean value of $dM(x, t)$. If we use superscripts 1 and 2 to denote the single or twin nature in the product density notation the mean and mean square number of individuals in the age group (x_1, x_2) are given by

(2.17) $$E\,\{[M(x_1, x_2; t)]^2\} = \int_{x_1}^{x_2} f_1^1(x, t)\, dx + 2 \int_{x_1}^{x_2} f_1^2(x, t)\, dx$$

$$(2.18)\ E\,\{[M(x_1, x_2; t)]^2\} = \int_{x_1}^{x_2} f_1^1(x, t)\, dx + 4 \int_{x_1}^{x_2} f_1^2(x, t)\, dx$$

$$+ \int_{x_1}^{x_2}\int_{x_1}^{x_2} f_2^{1,1}(x, y; t)\, dx\, dy$$

$$+ 2 \int_{x_1}^{x_2}\int_{x_1}^{x_2} f_2^{2,1}(x, y, t)\, dx\, dy$$

$$+ 2 \int_{x_1}^{x_2}\int_{x_1}^{x_2} f_2^{1,2}(x, y, t)\, dx\, dy$$

$$+ 4 \int_{x_1}^{x_2}\int_{x_1}^{x_2} f_2^{2,2}(x, y, t)\, dx\, dy.$$

Integral equations for these product densities have been derived by Ramakrishnan and Srinivasan [4] using arguments exactly similar to those employed
in the previous subsection. However, in the present case, we have to introduce
two subsidiary functions $K^1(x, t)$ and $K^2(x, t)$ where $K^i(x, t)\,dx$ denotes the
joint probability that at time t the primary is alive and an individual of i-th
type is found in the age group $(x, x + dx)$. The equations for the various
second order product densities are coupled and are capable of explicit solution
when the birth probabilities are constants.

2.3. BELLMAN-HARRIS REGENERATION POINT METHOD

Bellman and Harris [5] have dealt with the problem of finding the distribution of the total number of individuals generated by a single primary. Of
course, they have employed a different model of population growth in which
an object born at time 0 has random life-length at the end of which it is
replaced by a random number of similar objects of age 0. In the historic
paper published in 1948, they have introduced the method of regeneration
point which has wide applicability in different realms of physical science as
well. In this subsection, we shall adopt the method for a brief discussion of
our problem.

Let $\pi(n, x, x_0; t)$ be the probability of finding n individuals each having an
age greater than x at time t due to a primary of age x_0 at $t = 0$. Under the
assumptions stated in the beginning of this section, let us attempt to follow
the fate of the primary in the interval $(0, t)$. Somewhere between 0 and t, the
primary gives birth to a secondary or it does not, the probability of the latter
being

$$\exp\left[-\mu t - \int_0^t \lambda(u)\,du\right].$$

If it gives birth to a secondary at a particular instant, then the primary and
the secondary generate a population independently from that instant onward,
the point in the time axis being a point of regeneration. Using appropriate
probability weights, we find

$$(2.19) \qquad \pi(n, x, x_0; t) = \delta_{n1} \exp\left[-\mu t - \Lambda(t)\right]$$

$$+ \delta_{n0} \int_0^t \exp\left[-\mu y - \Lambda(y)\right] \mu \, dy$$

$$+ \sum_{n_1+n_2=n} \int_0^t \exp\left[-\mu y - \Lambda(y)\right] \lambda(y)$$

$$\pi(n_1, x, x_0 + y; t - y)\, \pi(n_2, x, 0; t - y)\, dy$$

where $\Lambda(y) = \int_0^y \lambda(u)\, du$. We can use the generating functional technique to write equation (2.19) in a convenient way.

Kendall [6] obtained the probability generating function for the total number of individuals generated in the special case when λ is a function of t and not of x. In a subsequent contribution, Kendall [7] dealt with another special case when λ is a constant and in this case he obtained the characteristic functional governing the total number of individuals present at time t. Davis [8] extended the results to include the distribution of the number of individuals above a certain age. These results are of great importance in the general theory of point processes since this is the only nontrivial situation where the characteristic functional (which contains all the statistical properties of the process in a "portmanteau" form) is available in an explicit manner.

We do not propose to go into the methods of solution of equation (2.19) and other similar equations, since these have been adequately discussed in the monograph of Bartlett [9].

3. Invariant Imbedding Technique

In the previous section, we have discussed the population growth on the basis of integral equations proposed by Kendall. At the time, Kendall proposed his integral equations, the regeneration point method which has since been generalized by Bellman and his collaborators [10] to a high degree of sophistication and is now well known by the name "invariant imbedding technique" was not much in vogue among population theorists. Accordingly, Kendall's procedure to evaluate the second moment involves the introduction and study of the subsidiary function $K(x, t)$ where $K(x, t)\, dx$ denotes the joint probability that a secondary of age between x and $x + dx$ is found at t and that the primary is alive at t. Subsequently, Ramakrishnan and Srinivasan [4] have extended Kendall's method to include twins in the age dependent theory of population growth. Their method consists in defining product densities of various orders of twins and singles and obtaining simultaneous integral equations for the product densities and certain subsidiary functions similar to the one introduced by Kendall. In this method, the number of such subsidiary functions increases with the order of the moment that is to be calculated. However, if we use the invariant imbedding technique, we need not introduce such subsidiary functions. The differential equation thus obtained connects the second order product densities directly to the first order product densities. Thus $f_1(x|x_0; t)$ satisfies (3.2) together with the initial condition:

$$(3.3) \qquad\qquad f_1(x|x_0; 0) = \delta(x - x_0)$$

where $\delta(x)$ is the Dirac-delta function.

In order to solve (3.2) under (3.3), we proceed along the characteristic curve $\psi = x_0 + t$ and obtain

$$\frac{\partial f_1(x|\psi - t; t)}{\partial t} + \mu(\psi - t) f_1(x|\psi - t; t) = \lambda(\psi - t) f_1(x|0; t)$$

which is only an ordinary differential equation and its solution can be written formally as

$$f_1(x|\psi - t; t) = \exp\left(-\int_0^t \mu(\psi - t')\, dt'\right)\left[\int_0^t \lambda(\psi - t')\right.$$

$$\left. (\exp\int_0^{t'} \mu(\psi - t'')\, dt'') f_1(x|0; t')\, dt' + F(\psi)\right].$$

Using the initial condition (3.3) we get

$$(3.4) \quad f_1(x|x_0; t) = \exp -\left(\int_0^t \mu(x_0 + t - t')\, dt'\right)$$

$$\left[\int_0^t \lambda(x_0 + t - t')\;\; f_1(x|0, t')\right.$$

$$\left. \left(\exp\int_0^t \mu(x_0 + t - t'')\, dt''\right) dt' + \delta(x - x_0 - t)\right].$$

This is a Volterra's integral equation of the third kind. But for some properly chosen $\lambda(x)$ and $\mu(x)$ the analytical solution of (3.4) becomes difficult. Even if we choose $\mu(x)$ to be proportional to x the integral equation is not easily susceptible to analytic solution even for a conveniently chosen $\lambda(x)$. However, in such cases, one can resort to the numerical solution. If one has in mind a population which is subject to some "risks" the probability of death is no more age dependent and one can safely take it to be a constant. With this restriction on $\mu(x)$, (3.4) becomes completely solvable for some realistic $\lambda(x)$. As an example, we shall deal with a situation in the process of cell division.

It is easy to see that $f_2(x, y | x_0; t)$ satisfies

(3.5) $f_2(x, y | x_0; 0) = 0$

$$f_2(x, y | x_0; t) = (1 - \lambda(x_0)\Delta - \mu(x_0)\Delta) f_2(x, y | x_0 + \Delta; t - \Delta)$$
$$+ \lambda(x_0)\Delta[f_2(x, y | 0; t)$$
$$+ f_2(x, y | x_0; t) + f_1(x | 0; t) f_1(y | 0; t)$$
$$+ f_1(y | 0; t) f_1(x | x_0; t)] + o(\Delta) \qquad (t > 0)$$

which in the limit as $\Delta \to 0$ becomes

(3.6) $$\frac{\partial f_2(x, y | x_0; t)}{\partial t} - \frac{\partial f_2(x, y | x_0; t)}{\partial x_0} + \mu(x_0) f_2(x, y | x_0; t)$$
$$= \lambda(x_0) f_2(x, y | 0; t) + \lambda(x_0)[f_1(x | 0; t) f_1(y | x_0; t)$$
$$+ f_1(y | 0; t) f_1(x | x_0; t)].$$

Proceeding, as earlier, along the characteristic curve $\psi = x_0 + t$ and "solving" the resulting ordinary differential equation, we obtain

(3.7) $$f_2(x, y | x_0; t) = \left[\exp - \int_0^t \mu(x_0 + t - t')\, dt' \right]$$

$$\int_0^t \lambda(x_0 + t - t') \left[\exp \int_0^{t'} \mu(x_0 + t - t'')\, dt'' \right]$$
$$[f_2(x, y | 0; t') + f_1(x | x_0 + t - t'; t')$$
$$f_1(y | 0; t') + f_1(y | x_0 + t - t'; t') f_1(x | x_0; t')]\, dt'$$

which, again, is a Volterra's integral equation of the third kind. As in the case of (3.4), here also we envisage the same difficulty in choosing $\mu(x)$ as age dependent. Hence we take $\mu(x)$ to be independent of x and obtain, in the next section, the solution of (3.7) for some age dependent birth process.

At this stage, we wish to remark that the second order product densities are connected only with the first order ones and thus there is no necessity for the introduction of auxiliary functions in contrast to Kendall's approach.

3.1. SOLUTION OF THE PROBLEM FOR AGE DEPENDENT BIRTH AND CONSTANT DEATH RATES

We shall presently solve the integral equations (3.4) and (3.7) choosing $\lambda(x) = xe^{-ax}$ where a is a constant and $\mu(x) = \mu$, a constant. Such a choice of $\lambda(x)$ is motivated by the ageing phenomenon in the cell division process of yeast cells [11].

While solving these equations it should be borne in mind that

(3.8) $f_1(x|x_0; t) = 0$ if $t < x$

and for definiteness if we assume $x < y$, then

(3.9) $f_2(x, y|x_0; t) = 0$ if $t < y$.

Now (3.4) becomes

$$(3.10) \qquad f_1(x|x_0; t)\, e^{\mu t} = \int_0^t (x_0 + t - t')\, e^{-a(x_0 + t - t')}$$

$$f_1(x|0; t')\, e^{\mu t'}\, dt' + \delta(x - x_0 - t).$$

In order to solve this equation, one can easily see that $f_1(x|0; t')$ occurring inside the integral is just $f_1(x|x_0; t)$ at $x_0 = 0$. Then $f_1(x|0; t)$ can be obtained by putting $x_0 = 0$ in (3.10) and solving the resulting integral equation

$$(3.11) \quad f_1(x|0; t) = \int_0^t (t - t')\, e^{-a(t-t') - \mu(t-t')} f_1(x|0, t')\, dt' + e^{-\mu t}\, \delta(x - t).$$

Using (3.8) one can easily observe that

(3.12) $f_1(x|0; t) = e^{-\mu t}\, \phi(t - x)$

where $\phi(u)$ satisfies

$$(3.13) \qquad \phi(u) = \int_0^u (u - t')\, e^{-a(u-t')}\, \phi(t')\, dt' + \delta(u)$$

which can be readily solved by the Laplace transform technique. Thus

(3.14) $\phi(u) = \delta(u) + \tfrac{1}{2} e^{-au} (e^u - e^{-u})\, H(u)$

where $H(u)$ is the Heaviside unit function. Now using (3.12) the solution of (3.10) can be seen to be

$$(3.15) \quad f_1(x|x_0; t)\, e^{\mu t} = \delta(x - x_0 - t) + \frac{e^{-ax_0}}{2} \left[e^{-(a-1)(t-x)} (x_0 + 1) \right.$$

$$\left. + e^{-(a+1)(t-x)} (x_0 - 1) \right] H(t - x).$$

Choosing $\lambda(x)$ and $\mu(x)$ as stated earlier in this section we see that (3.7) becomes

$$(3.16) \quad f_2(x, y|x_0; t)\, e^{\mu t} = \int_y^t (x_0 + t - t')f_2(x, y|0; t')\, e^{\mu t' - a(x_0 + t - t')}$$

$$+ \int_0^t (x_0 + t - t')[f_1(x|0; t')f_1(y|x_0 + t - t'; t')$$

$$+ f_1(y|0; t')f_1(x|x_0 - t + t'; t')]\, e^{\mu t' - a(x_0 + t - t')}\, dt'.$$

In order to solve this equation we only need to know $f_2(x, y|0; t)$ which precisely is $f_2(x, y|x_0; t)$ at $x_0 = 0$. Putting $x_0 = 0$ in (3.16) and using (3.12) and (3.15), it can be seen that

$$(3.17) \qquad\qquad f_2(x, y|0; t)\, e^{\mu t} = F(t - y, y - x)\, e^{-\mu y}$$

where $F(u, v)$ satisfies an integral equation in the variable u. An explicit expression for $f_2(x, y|x_0; t)$ has been obtained by Srinivasan and Koteswara Rao [12] and we do not propose to present the solution as it is fairly lengthy.

3.2. COMPARISON WITH KENDALL'S RESULTS

When $\lambda(x) = \lambda$ and $\mu(x) = \mu$ where λ and μ are constants, (3.4) becomes

$$(3.18) \quad f_1(x|x_0; t) = \delta(x - x_0 - t)\, e^{-\mu t} + \lambda \int_0^t f_1(x|0; t')\, e^{-\mu(t - t')}\, dt'$$

from which we obtain

$$(3.19) \qquad f_1(x|x_0; t) = \delta(x - x_0 - t)\, e^{-\mu t} + \lambda\, e^{-\mu t + \lambda(t - x)}\, H(t - x)$$

and (3.7) becomes

$$(3.20) \qquad f_2(x, y|x_0; t)\, e^{\mu t} = \lambda \int_0^t f_2(x, y|0; t')\, e^{\mu t'}\, dt'$$

$$+ \lambda \int_0^t [f_1(x|x_0 + t - t'; t')f_1(y|0; t')$$

$$+ f_1(y|x_0 + t - t'; t')f_1(x|0; t')]\, e^{\mu t'}\, dt'.$$

This can be solved using the method adopted in the previous section and thus we obtain the solution in the form given by (2.14).

In obtaining the solution we need not substitute the value of $f_2(x, y|0; t)$ in (3.20) and evaluate the integrals provided we observe the simplicity introduced in the process by making the birth rate a constant. In fact, $f_2(x, y|x_0; t)$ is the same as $f_2(x, y|0; t)$ provided the δ-term governing the primary is neglected as we are only interested in the secondaries. In a similar manner, higher order product densities can be obtained.

We wish to remark that a differential equation for the characteristic functional governing the age distribution can be written. Since it is not possible to obtain the explicit solution for the characteristic functional in closed form for general $\lambda(x)$, we have preferred to write down the differential equations for the product densities directly. There is another advantage arising from resorting to product densities. If initially the spectrum of age is defined statistically, the method brings out the fact that the process is Markovian in the sense that if the age dependent product densities specified at $t = 0$ up to any order are determined uniquely up to the same order for any $t > 0$.

4. Theory of Carcinogenesis

The statistical problem of carcinogenesis can be handled by the techniques of stochastical point processes. As is well known, the genesis of cancer in an individual is due to many causes apart from the direct administration of a carcinogen either in the form of a drug or by subjecting the individual to ionizing radiations. Still, experiments are being conducted on a large scale, to throw some more light on the cancerous tumors—how they form and develop—after administering a dose of carcinogen to guinea pigs. Various aspects of carcinogenesis are being studied which, in due course, may enable one to estimate the radiation hazards to the total world population from the fall-out from atomic weapon tests as well as from the increasing medical and industrial applications of atomic energy and ionizing radiation. The randomness in the genesis of cancer, apart from its pathology, constitutes an interesting study by itself and a number of feasible stochastic models have been proposed in recent years. In this section, we shall be concerned with a model proposed by Tucker [13] based on the results of the experiments of Polissar and Shimkin (see reference [13]). We shall discuss this model and present some results obtained by Srinivasan and Koteswara Rao [14] for an age dependent model. We shall also present a slightly different approach due to Kendall [15] and Neyman [16].

4.1. TUCKER'S MODEL

We shall sum up the main results of the experiments of Polissar and Shimkin on the basis of which the two stage theory of carcinogenesis emerges. A dose of the carcinogenic agent, urethan, was administered to each mouse in a collection of experimental mice. At certain intervals of time after administering the carcinogen a small group of mice was selected from the large group and their lungs were examined. In the observations made shortly after administering the carcinogen small distinct growths which were not cancerous have been noticed on the lungs. Polissar and Shimkin called these growths "hyperplastic foci." These hyperplastic foci increased both in number and size as time progressed. After a certain time the number of hyperplastic foci per mouse's lung began to decrease whereas their individual sizes became constant. Shortly after the hyperplastic foci were observed, tumors were observed on the lungs. For a while the tumors per mouse's lung increased in number and size but after a certain time the number became constant and the size alone was increasing. Thus Polissar and Shimkin considered the hyperplastic foci to be precursors of tumors and that some hyperplastic foci after attaining a certain age and size transformed into tumors. This is the two stage theory of carcinogenesis as proposed by Polissar and Shimkin. On the basis of this theory, Tucker proposed and studied a stochastic model under the following assumptions:

   (i) when a hyperplastic focus is first formed, it will be unobservable and after a certain time it can become either observable or disappear (die);

   (ii) an observable hyperplastic focus sometime after its formation can either transform into a tumor with a random number N of cells or die;

   (iii) the distribution of N is completely specified by its expectation;

   (iv) once a tumor is formed, it cannot undergo any further transformation other than increasing in size by increasing the number of cells it contained initially; and

   (v) the increase of cells in a tumor is a Poisson process with intensity $\rho(t)$, t denoting the time since the carcinogen is administered and *not* the time since the formation of the tumor, and for each tumor N will be independent of the number of cells subsequently added. The quantities of interest are the number of tumors and the number of cells at any time t after the administration of the carcinogen. Tucker assumed that the probability that an unobservable hyperplastic focus

   (i) will die during the interval $(t, t + h)$ is $\mu_0 h + o(h)$ and

   (ii) will become observable during the interval $(t, t + h)$ is $\nu_0 h + o(h)$ where μ_0 and ν_0 are non-negative constants.

In a similar manner, the probability that an observable hyperplastic focus dies in an interval $(t, t + h)$ is $\mu h + o(h)$ while the probability of its getting transformed into a tumor during that interval is $\nu(t)h + o(h)$ where ν is non-negative function of t. We further assume that μ and ν depend only on the carcinogen and the animal and not on the dosage. On the basis of this model, Tucker obtained some explicit results by first obtaining the joint probability generating function of the number of observable hyperplastic foci, the number of twins, and the number of cells in all the tumors on the lungs of the guinea pig. As Tucker has pointed out, the model would be more realistic if the various transition probabilities could be made age dependent. This has been achieved by Srinivasan and Koteswara Rao [14] who used the product density technique to obtain information on the mean and fluctuation about the mean number of cells. In the next few subsections we shall summarize the main results of this investigation.

4.2. AN AGE DEPENDENT MODEL

The model is based on the following assumptions:

(i) $f(t)$ is the "feeding function" determined jointly by the animal, the dosage, and the type of carcinogen. The probability that the carcinogen causes the formation of one unobservable hyperplastic focus during $(t, t + dt)$ is $f(t) dt + o(dt)$ where t is the time measured from the instant the carcinogen is administered;

(ii) the unobservable hyperplastic focus can either become an observable hyperplastic focus or disappear (die). All unobservable hyperplastic foci act independently of each other with regard to becoming observable or dying.

The probability that an unobservable hyperplastic focus of age x at time t becomes observable in the interval $(t, t + dt)$ is $v_0(x) dt + o(dt)$.

The probability that an unobservable hyperplastic focus of age x at time t dies in the interval $(t, t + dt)$ is $\mu_0(x) dt + o(dt)$.

The probability that more than one unobservable hyperplastic focus either becomes observable or dies in the interval $(t, t + dt)$ is $o(dt)$.

If $0 < t_1 < t_2 < t_3$ the number of hyperplastic foci formed because of the carcinogen in the two intervals (t_1, t_2) and (t_2, t_3) are independent.

(iii) The probability that an observable hyperplastic focus of age x at time t transforms into a tumor in the interval $(t, t + dt)$ is $v(x) dt + o(dt)$.

The probability that an observable hyperplastic focus of age x dies in the interval $(t, t + dt)$ is $\mu(x) dt + o(dt)$.

The probability that more than one observable hyperplastic focus either becomes a tumor or dies in the interval $(t, t + dt)$ is $o(dt)$. The numbers of tumors formed in two different intervals are independent.

(iv) Every newborn tumor will consist of N cells, N being a random number the distribution of which is one that is completely specified by its expectation.

(v) The cells in a tumor will increase in a Poissonian way with intensity $\rho(t)$. N will be independent of the number of cells subsequently added.

4.3. PRODUCT DENSITY APPROACH

Let $f_1^O(x, t)$, $f_1^U(x, t)$ denote the product densities of first order of the observable and unobservable hyperplastic foci. If $f_1^T(t)$ is the first order production product density* (see Section 4 of Chapter 4) of tumors, using the fact that $f_1^T(t)\, dt$ is a probability magnitude, we obtain

$$(4.1) \qquad f_1^T(t) = \int_0^t f_1^O(u, t)\, v(u)\, du$$

$$f_1^O(u, t) = \int_0^{t-u} f_1^U(p, t-u)\, v_0(p) \left[\exp - \int_0^U [(\mu(t') + v(t'))\, dt'] \right] dp$$

$$f_1^U(p, t-u) = f(t-u-p) \exp \left\{ - \int_0^p [\mu_0(t') + v_0(t')]\, dt' \right\}$$

and hence

$$(4.2) \quad f_1^T(t) = \int_0^t v(u) \left[\exp - \int_0^u [\mu(t') + v(t')]\, dt' \right] du$$

$$\int_0^{t-u} f(t-u-p)\, v_0(p) \left\{ \exp - \int_0^p [\mu_0(t') + v_0(t')]\, dt' \right\} dp.$$

Let $f_2^O(x, y; t)$ and $f_2^U(x, y; t)$ denote the second order product densities for observable and unobservable hyperplastic foci. Let $f_2^T(t_1, t_2)$ denote the second order production product density for tumors.

In view of the fact that the formation of a hyperplastic focus depends only on the intensity of the carcinogen and not on the presence of other hyperplastic foci, the higher order product densities of tumors can always be expressed as a product of first order product densities. For example, the second order product density for tumors is given by

$$(4.3) \qquad\qquad f_2^T(t_1, t_2) = f_1^T(t_1) f_1^T(t_2).$$

---

* Since tumors do not undergo any transformation other than increasing in their individual sizes we are only interested in the production product densities.

Similarly

(4.4)
$$f_2^O(x, y, t) = f_1^O(x, t) f_1^O(y, t)$$
$$f_2^U(x, y; t) = f_1^U(x, t) f_1^U(y, t).$$

We shall use the random variables $Y(t)$ and $Z(t)$ to denote respectively the number of tumors and cells existing at time t after administering the carcinogen. Using the moment formula (2.12) of Chapter 3 and the fact that the higher order product densities are products of the first order product densities for tumors we obtain

(4.5)
$$\overline{[Y(t)]^r} = \sum_{s=1}^{n} C_s^r \left[\int_0^t f_1^T(t')\, dt' \right]^s$$

where C_s^r is the usual weight coefficient representing the number of $(r - s)$-fold degeneracies in a product of r terms.

Since $dY(t)$ denotes the number of tumors formed in the interval $(t, t + dt)$ we get

(4.6)
$$\overline{dY(t)} = f_1^T(t)\, dt.$$

In view of the assumptions made on the development of the number of cells in a tumor we find the random variable $Z(t)$ to be given by

(4.7)
$$Z(t) = \int_0^t dY(t')[N + n(t - t')]$$

where $n(t - t')$ is the number of cells subsequently added to the initial number in time $t - t'$, where t' is the time of formation of the tumor. The assumption that the number of cells increases in a Poisson way is motivated by Tucker's paper. However, it may be worthwhile to assume that the increase of cells follows a Furry distribution. Our results will hold good in both the cases as $n(t - t')$ has to be evaluated using the desired distribution. We will be able to get the required results using (4.6) and (4.7). Thus

(4.8)
$$\overline{Z(t)} = \int_0^t \overline{dY(t - t')}\, (\bar{N} + \overline{n(t - t')})$$

(4.9)
$$\sigma^2[Z(t)] = \int_0^t \int_0^t \overline{dY(t_1')\, dY(t_2')\, (N + n(t - t_1')(N + n(t - t_2'))} - \overline{[Z(t)]^2}.$$

Observing that

$$Z(t) - NY(t) = \int_0^t dY(t')\, n(t - t')$$

we find the r-th moment of the number of subsequently added cells to be given by

$$\overline{[Z(t) - NY(t)]^r} = \int_0^t \int_0^t \cdots \int_0^t \overline{dY(t_1)\, dY(t_2) \cdots dY(t_r)}$$

$$\overline{n(t - t_1)\, n(t - t_2) \cdots n(t - t_r)}$$

$$= \sum_{h=1}^{r} \sum_i C_h^r(i) \int_0^t \int_0^t \cdots \int_0^t f_h^T(t_1, t_2, \ldots, t_h)$$

$$\overline{[n(t - t_1)]^{l_1} [n(t - t_2)]^{l_2} \cdots [n(t - t_h)]^{l_h}}$$

where $C_h^r(i)$ denotes the number of ways of obtaining a typical complexion i, the complexion being characterized by the set of number $l_1, l_2, \ldots, l_h$ such that $l_1 + l_2 + \cdots + l_h = r$.

In view of the assumption that the cells increase in a non-homogeneous Poisson manner with intensity $\rho(t)$ and that the increase in the cells in a particular tumor is independent of that in any other tumor and that the higher order product densities for tumors are only products of the first order product densities, we get

(4.10) $\quad \overline{[Z(t) - NY(t)]^r}$

$$= \sum \sum_{h=1} C_h^r(i) \prod_{i\ m=1}^{h} \left\{ \int_0^t f_1^T(t_m) \left[\sum_{k_m=1}^{1m} C_{k_m}^{1m} \left(\int_{t_m}^t \rho(v)dv \right)^{k_m} \right] dt_m \right\}$$

This process of increase in the number of cells can be viewed from the point of random processes associated with random points on a line, observing the time axis as the line and the times of formation of tumors as the random points on it and the development of the cells in each tumor as the random processes associated with them (see reference [17]).

The covariance of $Y(t)$ and $Z(t)$ is given by

$$\operatorname{cov}[Y(t), Z(t)] = \int_0^t \int_0^t \overline{dY(t_1')\, dY(t_2')\, [N + n(t - t_2')]} - \overline{Y(t)}\,\overline{Z(t)}.$$

In view of (4.3) we find that

(4.11) $\qquad\qquad\qquad \operatorname{cov}(Y(t), Z(t)) = \overline{Z(t)}.$

The results (4.8), (4.9) and (4.11) also agree with the results of Tucker [see equations (5.8), (5.11), and (5.14) of his paper] for the special case treated by him. In fact these results are given in a more general form to cover the case when the transition probabilities are age dependent.

We observe that (4.10) holds good even when the cells increase according to a Furry law (see Section 2 of Chapter 1) if we replace C_{km}^{lm} by $C_{km}^{lm} k_m!$. This is due to the fact that in the case of a Furry process the n-th order product density of events is given by a simple law.

We wish to remark, at this stage, that if we are interested in the mutation process introduced by Tucker (see Section 4 of his paper), following the above line of approach it is easy to write down the product densities of various orders for a particle of the i-th type even when the transition probabilities are age dependent. Using the properties of the product densities, we can obtain the moments of the number of particles of the i-th type.

4.4. KENDALL-NEYMAN APPROACH TO CARCINOGENESIS

So far, we have dealt with a particular model of carcinogenesis motivated by the experiments of Polissa and Shimkin. However, there are other models of carcinogenesis worthy of investigation. In 1959, Neyman (see Neyman and Scott [16]) has observed that some calculations of Kendall [15] relating to phenotypically delayed mutations can be adopted from a rough quantitative model for the theory of carcinogenesis. Kendall [18] formulated the problem in terms of a birth and death process and obtained estimates for the distribution of the cells constituting a cancerous growth. This model has been further studied by Waugh [19] who has apparently generalized some of the results of Kendall.

Following Kendall, let us first outline the biological situation and the model introduced to represent it. To start with, we have a large population of normal cells subject to carcinogenic action usually of two kinds: a "background" which is always present and an enhancement due to the experimental conditions. We assume that the joint effect is to transform normal cells into "gray cells" (coined by Kendall), the number of gray cells so produced being governed by a Poisson law characterized by the time dependent parameter $f(t)$, with t being measured from the time of commencement of the experiment. Each gray cell generates a clone, the individuals in which multiply with birth and death rates λ and μ respectively, μ being greater than λ so that the gray clones will become extinct over a sufficiently long period of time. Apart from this, the gray cells are capable of being transformed into second order mutants called black cells, the transformation proceeding through either of the two following alternatives:

(A) a gray cell gets transformed into a single gray cell and a black cell, this being independent of the birth and death process controlling the growth of the clone to which the gray cell belongs;

(B) second order mutations occur at epochs of cell division; their effect is to convert one of the two fission products into a black cell.

Once a black cell is formed, it generates a black cell clone developing according to a birth-and-death process, the birth rate L being greater than the death rate M. Thus the malignant growths consisting of black clones are identified with supercritical birth-and-death processes while the benign growths consisting of clones of gray cells are identified with subcritical processes.

Kendall has assumed that for a gray cell present at t, there is a chance $\lambda \, dt$ that it splits into two gray cells during $(t, t + dt)$ and that there is a chance $\mu \, dt$ that it will "die" (become a normal cell back again) during the same period. The probability with which a gray cell undergoes a second order transition in $(t, t + dt)$ is assumed to be $\nu \, dt$ and ν is taken to be a constant time $f(t)$ where $f(t)$ is the function used to measure the intensity of the carcinogenic action at various times. As we have mentioned in the introductory remarks above, the main object is to estimate the probability frequency function governing the number and size of the black clones. Kendall has completely solved the problem by obtaining the double generating function governing the distribution of the number of gray cells and black clones generated by a single gray cell. He has also dealt with the general problem taking into account the formation of gray cells at different points on the t-axis. Waugh [19] has obtained the double generating function in a more general situation where the probabilities λ and μ are age dependent. However, the calculations are restricted to the distribution of gray cells and black clones generated by a single gray cell at $t = 0$. The general problem which takes into account the formation of independent gray cells at different points on the t-axis still remains open. We wish to emphasize that the product density technique can be readily employed as in Section 4.3 and an estimate of the first few moments of the distribution of the number of clones can be obtained. This is the decided advantage of the product density technique in such non-Markovian situations.

REFERENCES

1. H. W. Watson and F. Galton, *Educational Times*, **19** (1873), 103.
2. Symposium on Stochastic Processes, *J. Roy. Statist. Soc.*, **B 11** (1949), 150.
3. D. G. Kendall, *J. Roy. Stat. Soc.*, **B 11** (1949), 230.

4. A. Ramakrishnan and S. K. Srinivasan, *Bull. Math. Bio. Phys.*, **20** (1958), 289.
5. R. E. Bellman and T. E. Harris, *Proc. Nat. Acad. Sci. U.S.A.*, **34** (1948), 601.
6. D. G. Kendall, *Ann. Math. Statist.*, **19** (1948), 1.
7. D. G. Kendall, *J. Roy. Statist. Soc.*, **B 12** (1950), 278.
8. A. W. Davis, *Biometrika*, **51** (1964), 245.
9. M. S. Bartlett, *An Introduction to Stochastic Processes*, Cambridge University Press, 1955.
10. R. E. Bellman, R. E. Kalaba, and G. M. Wing, *J. Math. Phys.*, **1** (1960), 280.
11. R. K. Mortimer and J. R. Johnston, *Nature*, **183** (1959), 1751.
12. S. K. Srinivasan and N. V. Koteswara Rao, *J. Math. Analy. Appl.*, **21** (1968), 43.
13. H. G. Tucker, *Proc. Fourth Berkeley Symp. Math. Statist.*, Vol. IV (1961), 387.
14. S. K. Srinivasan and N. V. Koteswara Rao, *I.I.T. Report*, No. 29, 18th Aug. 1967.
15. D. G. Kendall, *Ann. Inst. H. Poincaré*, **13** (1952), 43.
16. J. Neyman and E. L. Scott, *Science*, **130** (1959), 303.
17. S. K. Srinivasan and K. S. S. Iyer, *Zastosawania Matematyki*, **8** (1966), 221.
18. D. G. Kendall, *Biometrika*, **47** (1960), 13.
19. W. A. O'N Waugh, *Proc. Fourth Berkeley Symp. Math. Statist.*, Vol. IV (1961), 405.

AUTHOR INDEX

Numbers in parentheses indicate the numbers of the references when these are cited in the text without the name of the author. Numbers set in *italics* designate the page numbers on which the complete literature citation is given.

SUBJECT INDEX